Synthesis Lectures on Mechanical Engineering

This series publishes short books in mechanical engineering (ME), the engineering branch that combines engineering, physics and mathematics principles with materials science to design, analyze, manufacture, and maintain mechanical systems. It involves the production and usage of heat and mechanical power for the design, production and operation of machines and tools. This series publishes within all areas of ME and follows the ASME technical division categories.

Seiichi Nomura

Complex Variables
for Engineers
with Mathematica

Springer

Seiichi Nomura
University of Texas at Arlington
Arlington, TX, USA

ISSN 2573-3168   ISSN 2573-3176 (electronic)
Synthesis Lectures on Mechanical Engineering
ISBN 978-3-031-13069-4   ISBN 978-3-031-13067-0 (eBook)
https://doi.org/10.1007/978-3-031-13067-0

This Springer imprint is published by the registered company Springer Nature Switzerland AG
The registered company address is: Gewerbestrasse 11, 6330 Cham, Switzerland

Preface

Complex variable theory is one of the most important subjects in mathematics, and many excellent textbooks have been written. Complex variable theory has not changed over centuries. As this is a classical branch in mathematics, there seems hardly any need for yet another book on complex variables.

Complex variable theory is attractive for engineers as it offers elegant approaches for certain types of differential equations in engineering including heat transfer, solid mechanics and fluid mechanics. However, a gap exists between books written by mathematicians and books written by engineers in their specific fields. Naturally, mathematicians tend to emphasize rigorousness and consistency while less emphasizing applications. On the other hand, books written by engineers often jump directly to the specific topics assuming that the readers already have sufficient background of complex variables and the pathway from theory to application is not clearly elucidated. Therefore, those who want to learn how complex variable theory is utilized in their fields are often frustrated when necessary background materials are not provided.

One of the motivations for writing this book is to close the gap above such that a smooth transition from basic theory to application is accomplished. Although it is not possible to cover all the topics in engineering exhaustively, the readers can at least find the logic of how and why complex variables are effective for some of the engineering problems.

Another motivation for writing this book is to demonstrate that the readers can take advantage of a computer algebra system, *Mathematica*, to facilitate tedious algebra and visualize complex functions so that they can focus on principles instead of spending endless hours on algebra by hand. Unlike numerical tools such as MATLAB and FORTRAN, *Mathematica* can expand, differentiate, and integrate complex-valued functions symbolically. *Mathematica* can be used as a stand-alone symbolic calculator or a programming tool using the Wolfram Language. If *Mathematica* is not available locally, Wolfram Cloud Basic can be used online as a free service to execute *Mathematica* statements.

This book is suitable for upper-level undergraduate students or graduate students in STEM who are interested in the expeditious application of complex variables in their selected fields. The only prerequisite is the undergraduate level of calculus.

The book consists of six chapters and Appendix. Chapters 1–5 discuss the fundamentals of complex variables in a manner less rigorous but just necessary for the readers to prepare for engineering application, which is discussed in Chap. 6. Chapter 1 is an introduction to complex variables and various complex functions such as exponential, logarithmic and trigonometric functions transitioning from their real counterparts. Chapter 2 examines differentiability of complex-valued functions known as the Cauchy-Riemann equations and the concept of analytic functions that are the major players in complex variables. Chapter 3 discusses integral calculus of complex functions known as the Cauchy theorem and Cauchy's integral formula. The fundamental theorem of algebra can be proven using the Liouville theorem directly from Cauchy's integral formula. Chapter 4 is a series expansion of complex functions known as the Taylor and Laurent series. The concept of analytic continuation is used to show a stunning identity $1 + 2 + 3 + \ldots = -\frac{1}{12}$ known as the Ramanujan Summation. Chapter 5 discusses the residue theorem and its applications to improper integrals. This topic used to be one of the highlights of complex variables, but it is of less importance today as symbolic software including *Mathematica* can automatically carry out many integrations. Chapter 6 is the most important chapter in the book and shows how complex variables can be applied to real engineering problems. Topics are selected from heat transfer, solid mechanics and fluid mechanics and conformal mapping as well as a series expansion method are explained. Appendix is an introduction to *Mathematica* essentials. Each chapter has sample *Mathematica* codes to help understand the subject. Just like the best way to learn a programming language is to actually type the program yourself, I suggest that the readers enter the *Mathematica* commands in the book themselves to feel how complex variables work in action.

I cannot deny that this book was influenced by Prof. Michael D. Greenberg of the University of Delaware and his popular textbook [8]. His unique style in the lecture and book lead me to believe that engineering is, after all, an interpretation of mathematics in chosen fields. I would like to acknowledge my gratitude to Prof. Greenberg. I would also like to thank my colleagues, Dr. Erian Armanios and Dr. Kathy Hays-Stang, for their willingness to help, and Paul Petralia of Springer Nature for his support and encouragement.

Arlington, TX, USA   Seiichi Nomura

Contents

1 Functions of Complex Variables 1
  1.1 Complex Numbers 1
  1.2 Complex Plane 4
  1.3 Complex-Valued Functions 8
      1.3.1 Exponential Function, e^z 9
      1.3.2 Trigonometric Functions 10
      1.3.3 Logarithmic Function, $\log z$ 13
      1.3.4 Branch Cut and Branch Points 15
  1.4 Numerics 19
  1.5 Problems 21

2 Calculus of Functions of Complex Variables 23
  2.1 Differentiability (Cauchy-Riemann Equations) 23
      2.1.1 Cauchy-Riemann Equations 24
      2.1.2 Alternative Form of Cauchy-Riemann Equations 27
      2.1.3 Harmonic Functions 28
      2.1.4 Uniqueness of Analytic Functions 30
  2.2 Problems 32

3 Integrations of Functions of Complex Variables 35
  3.1 Integral Calculus 35
  3.2 Cauchy's Theorem 38
      3.2.1 Morera's Theorem 41
  3.3 Cauchy's Integral Formula 42
      3.3.1 Contour Integral of z^n 42
      3.3.2 Cauchy's Integral Formula 43
      3.3.3 Generalized Cauchy's Integral Formula 44
      3.3.4 Liouville's Theorem 48
      3.3.5 Fundamental Theorem of Algebra 51
  3.4 Problems 52

4 Series of Complex Variable Functions 53
   4.1   Taylor Series 53
       4.1.1 Taylor Series of $f(z)$ About $z = a$ 54
       4.1.2 Analytic Continuation 60
       4.1.3 Can We Prove $1 + 2 + 3 + 4 + \cdots = -\frac{1}{12}$? 63
   4.2   Laurent Series 64
   4.3   *Mathematica Code* 72
   4.4   Problems 73

5 Residues 75
   5.1   Types of Singularities 75
   5.2   Residues 78
       5.2.1 Definition of Residues 78
       5.2.2 Calculation of Residues 78
   5.3   Residue Theorem 82
       5.3.1 Residue at Infinity 83
   5.4   Application of Residue Theorem to Certain Integrals 86
       5.4.1 First Type 87
       5.4.2 Second Type 89
       5.4.3 Third Type 97
       5.4.4 *Mathematica* Code 101
   5.5   Problems 103

6 Applications to Engineering Problems 105
   6.1   Conformal Mapping 105
       6.1.1 Solving Laplace Equation by Conformal Mapping 107
       6.1.2 Bilinear (Möbius) Transformation 112
   6.2   General Solution to Laplace Equation ($\Delta\phi(x, y) = 0$) 112
   6.3   General Solution to Bi-harmonic Equation ($\Delta\Delta\phi(x, y) = 0$) 114
   6.4   Heat Conduction 116
   6.5   Solid Mechanics 126
   6.6   Fluid Mechanics 137
   6.7   Problems 150

Appendix: Introduction to *Mathematica* 153

References 169

Index 171

Functions of Complex Variables

1

1.1 Complex Numbers

A formal way to define a complex number, z, is to associate z with a pair of real numbers, x and y, as

$$z : (x, y). \tag{1.1}$$

Addition, subtraction and multiplication between two complex numbers, z_1 and z_2, expressed as

$$z_1 : (x_1, y_1), \quad z_2 : (x_2, y_2) \tag{1.2}$$

can be defined as

$$z_1 \pm z_2 : (x_1 \pm x_2, y_1 \pm y_2), \tag{1.3}$$

$$z_1 \cdot z_2 : (x_1 x_2 - y_1 y_2, x_1 y_2 + x_2 y_1). \tag{1.4}$$

Division of z_1 by z_2 is done by finding z that satisfies

$$z_1 = z_2 \cdot z. \tag{1.5}$$

By defining $z : (x, y)$, Eq. (1.5) can be written as

$$(x_1, y_1) = (x_2 x - y_2 y, x_2 y + y_2 x). \tag{1.6}$$

Equation (1.6) can be solved for x and y as

$$x = \frac{x_1 x_2 + y_1 y_2}{x_2^2 + y_2^2}, \quad y = \frac{x_2 y_1 - x_1 y_2}{x_2^2 + y_2^2}, \tag{1.7}$$

© The Author(s), under exclusive license to Springer Nature Switzerland AG 2022
S. Nomura, *Complex Variables for Engineers with Mathematica*,
Synthesis Lectures on Mechanical Engineering,
https://doi.org/10.1007/978-3-031-13067-0_1

or equivalently,

$$\frac{z_1}{z_2} : \left(\frac{x_1 x_2 + y_1 y_2}{x_2^2 + y_2^2}, \frac{x_2 y_1 - x_1 y_2}{x_2^2 + y_2^2} \right). \tag{1.8}$$

With this notation, an *imaginary number* denoted as i can be defined as

$$i : (0, 1). \tag{1.9}$$

It follows that

$$i \cdot i : (0, 1) \cdot (0, 1) = (-1, 0). \tag{1.10}$$

If we write Eq. (1.1) as

$$z = x + yi, \tag{1.11}$$

Equation (1.10) can be written as

$$i \cdot i = -1. \tag{1.12}$$

This way, all the algebraic operations among complex numbers can be performed as if they were real numbers except that every occurrence of i^2 must be replaced by -1. Thus, we can handle complex numbers just like real numbers plus i without mentioning that $i = \sqrt{-1}$.

Instead of using the notation of Eq. (1.1), we will use the notation of Eq. (1.11) from now on. We can state that a complex number, z, is a combination of two real numbers, x and y, through a symbol, i, called the imaginary number in the form of

$$z = x + yi, \tag{1.13}$$

where x is called the real part of z and is denoted as $\Re(z)$, and y is called the imaginary part of z denoted as $\Im(z)$.

The complex conjugate of $z = x + yi$ is denoted as $\bar{z}$ and defined as

$$\bar{z} = x - yi, \tag{1.14}$$

i.e., changing the sign of the imaginary part of z.

The following identities are important:

$$\overline{z_1 + z_2} = \overline{z_1} + \overline{z_2}, \tag{1.15}$$

$$\overline{z_1 z_2} = \overline{z_1} \, \overline{z_2}, \tag{1.16}$$

$$\overline{\left(\frac{z_1}{z_2} \right)} = \frac{\overline{z_1}}{\overline{z_2}}. \tag{1.17}$$

The proof is straightforward. For instance, Eq. (1.17) can be proven from Eq. (1.8)

$$\overline{\left(\frac{z_1}{z_2}\right)} = \overline{\left(\frac{x_1 x_2 + y_1 y_2}{x_2^2 + y_2^2} + i\frac{x_2 y_1 - x_1 y_2}{x_2^2 + y_2^2}\right)}$$
$$= \frac{x_1 x_2 + y_1 y_2}{x_2^2 + y_2^2} - i\frac{x_2 y_1 - x_1 y_2}{x_2^2 + y_2^2}. \tag{1.18}$$

On the other hand,

$$\frac{\overline{z_1}}{\overline{z_2}} = \frac{x_1 - y_1 i}{x_2 - y_2 i}$$
$$= \frac{x_1 x_2 + y_1 y_2}{x_2^2 + y_2^2} - i\frac{x_2 y_1 - x_1 y_2}{x_2^2 + y_2^2}. \tag{1.19}$$

The magnitude of $z = x + yi$, also called the absolute value of z, denoted as $|z|$, is defined by

$$|z| \equiv \sqrt{x^2 + y^2}. \tag{1.20}$$

Note that

$$z\bar{z} = |z|^2. \tag{1.21}$$

Mathematica **Programming**

All of scientific/engineering application software today supports complex numbers but applications that can handle symbolic manipulation such as *Mathematica* and Maple have advantages over numerical applications represented by MATLAB in that exact algebra/calculus can be carried out without numerical errors.

This book adopts *Mathematica* to handle complex numbers because of its capabilities of processing complex numbers analytically. You can let the computer do algebra and analysis and concentrate on the principles. You can access Wolfram Cloud (https://www.wolframcloud.com/) for most operations of *Mathematica* with a basic free subscription plan which enable a Notebook interface similar to the desktop experience. The Appendix introduces the fundamental operations of *Mathematica*.

In *Mathematica*, a capitalized *I* is used to represent the imaginary number, $\sqrt{-1}$.[1]

```
In[ ]:= I^2
Out[ ]= -1

In[ ]:= I^101
Out[ ]= i
```

A complex number, $z = x + yi$, can be entered as

```
In[ ]:= z = x + y I
Out[ ]= x + i y
```

---

[1] You can also enter the imaginary number by pressing Esc, i,i, Esc keys.

Note that a space between y and I can be used for multiplication. An asterisk (*) can be also used for multiplication. However, entering z ^ 3 does not automatically expand $(x + yi)^3$ into the real part and the imaginary part. If you enter z^3,

In[ ]:= **z^3**

Out[ ]= $(x + i\, y)^3$

Mathematica simply returns $(x + yi)^3$ because it assumes by default that variables, x and y, could be also complex numbers and cannot separate real and imaginary parts. If you want to see the real part and the imaginary part of $(x + yi)^3$ separately, use the ComplexExpand[ ] function which assumes that all the symbols are real as

In[ ]:= **ComplexExpand[z^3]**

Out[ ]= $x^3 - 3\,x\,y^2 + i\,(3\,x^2\,y - y^3)$

To extract the real and imaginary parts from a complex number, use the Re[ ] (real) and Im[ ] (imaginary) functions followed by the ComplexExpand[ ] function as

In[ ]:= **ComplexExpand[z^3]**

Out[ ]= $x^3 - 3\,x\,y^2 + i\,(3\,x^2\,y - y^3)$

In[ ]:= **Im[z^3]**

Out[ ]= $\mathrm{Im}\big[(x + i\, y)^3\big]$

In[ ]:= **ComplexExpand[Re[z^3]]**

Out[ ]= $x^3 - 3\,x\,y^2$

In[ ]:= **ComplexExpand[Im[z^3]]**

Out[ ]= $3\,x^2\,y - y^3$

The ComplexExpand function works on a division as well.

In[ ]:= **ComplexExpand[1/z^3]**

Out[ ]= $\dfrac{x^3}{(x^2+y^2)^3} - \dfrac{3\,x\,y^2}{(x^2+y^2)^3} + i\left(-\dfrac{3\,x^2\,y}{(x^2+y^2)^3} + \dfrac{y^3}{(x^2+y^2)^3}\right)$

1.2   Complex Plane

Since a complex number, $z = x + yi$, is, after all, a set of two real numbers linked through i, it can be visualized by associating z with a point (x, y) in a 2-D plane. A 2-D plane where each point represents a complex number is called a complex plane. In Fig. 1.1, the horizontal axis is called the real axis and the vertical axis is called the imaginary axis.

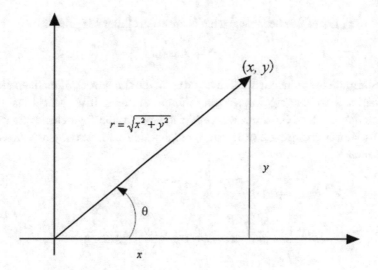

Fig. 1.1 Complex plane and polar form

The magnitude of the vector from $(0, 0)$ to (x, y) in Fig. 1.1 is the absolute value of z as

$$|z| = \sqrt{x^2 + y^2}$$
$$= \sqrt{z\bar{z}}. \tag{1.22}$$

The angle, θ, in Fig. 1.1 is called the argument of z and is denoted as

$$\arg z = \theta. \tag{1.23}$$

We can express z in terms of (r, θ) from (x, y) as

$$z = x + yi$$
$$= \sqrt{x^2 + y^2}\left(\frac{x}{\sqrt{x^2 + y^2}} + i\frac{y}{\sqrt{x^2 + y^2}}\right)$$
$$= r(\cos\theta + i\sin\theta)$$
$$= re^{i\theta}, \tag{1.24}$$

where Euler's formula,

$$e^{i\theta} = \cos\theta + i\sin\theta, \tag{1.25}$$

was used. This relationship known as Euler's formula warrants substantial discussion. However, for now, Eq. (1.25) can be understood as the definition of $e^{i\theta}$.

When $\theta = \pi$, Eq. (1.25) becomes[2] what is known as Euler's identity as

$$e^{i\pi} + 1 = 0. \tag{1.26}$$

Euler's formula is not inconsistent in any way in the framework of mathematics and we can even derive it without rigor in several different manners. If we accept that the Taylor series for complex-valued functions works the same way as the Taylor series for real-valued functions by simply changing x (real) to z (complex), the complex Taylor series of $e^{i\theta}$ can be expressed as

$$
\begin{aligned}
e^{i\theta} &= 1 + (i\theta) + \frac{(i\theta)^2}{2!} + \frac{(i\theta)^3}{3!} + \frac{(i\theta)^4}{4!} + \cdots \\
&= \left(1 - \frac{\theta^2}{2!} + \frac{\theta^4}{4!} - \cdots\right) + i\left(\theta - \frac{\theta^3}{3!} + \frac{\theta^5}{5!} - \cdots\right) \\
&= \cos\theta + i\sin\theta, \tag{1.27}
\end{aligned}
$$

which is the same as Eq. (1.25).

Another way of deriving Euler's formula without rigor is to define y as

$$y \equiv \cos\theta + i\sin\theta. \tag{1.28}$$

Differentiation of y with respect to θ yields

$$
\begin{aligned}
\frac{dy}{d\theta} &= -\sin\theta + i\cos\theta \\
&= i^2\sin\theta + i\cos\theta \\
&= i(\cos\theta + i\sin\theta) \\
&= i\,y. \tag{1.29}
\end{aligned}
$$

Solving $\frac{dy}{d\theta} = i\,y$ with $y(0) = 1$, we have

$$y = e^{i\theta}, \tag{1.30}$$

which is Eq. (1.25). Euler's formula implies that the exponential functions and the trigonometric functions are interchangeable through i.

If we assume that $e^{i\theta}$ behaves similar to e^x, we can state

$$\left(e^{i\theta}\right)^n = e^{in\theta}, \tag{1.31}$$

where n is an integer. Equation (1.31) can be written as

---

[2] Arguably, this identity is cited as an example of deep mathematical beauty as each component represents each of the branch of mathematics, i.e., e for analysis, π for geometry, "1" and "0" for algebra as the units of multiplication and addition [5].

$$(\cos \theta + i \sin \theta)^n = (\cos n\theta + i \sin n\theta), \qquad (1.32)$$

which is known as de Moivre's formula and is often used to reduce $\cos n\theta$ and $\sin n\theta$ in terms of $\cos \theta$ and $\sin \theta$. For example, for $n = 3$, Eq. (1.32) is written as

$$(\cos \theta + i \sin \theta)^3 = \cos 3\theta + i \sin 3\theta. \qquad (1.33)$$

The left-hand side of Eq. (1.33) is expanded as

$$\cos^3 \theta - 3 \sin^2 \theta \cos \theta + i \left(3 \sin \theta \cos^2 \theta - \sin^3 \theta\right). \qquad (1.34)$$

Therefore, by comparing the real and imaginary parts, the following relationships are derived:

$$\cos 3\theta = \cos^3 \theta - 3 \sin^2 \theta \cos \theta, \qquad (1.35)$$
$$\sin 3\theta = 3 \sin \theta \cos^2 \theta - \sin^3 \theta. \qquad (1.36)$$

In *Mathematica*, the above derivation can be achieved by using the `ComplexExpand[]` function.

```
In[ ]:= z = (Cos[th] + I Sin[th])^3
Out[ ]= (Cos[th] + i Sin[th])^3
```

```
In[ ]:= ComplexExpand[z]
Out[ ]= Cos[th]^3 - 3 Cos[th] Sin[th]^2 + i (3 Cos[th]^2 Sin[th] - Sin[th]^3)
```

```
In[ ]:= ComplexExpand[Re[z]]
Out[ ]= Cos[th]^3 - 3 Cos[th] Sin[th]^2
```

```
In[ ]:= ComplexExpand[Im[z]]
Out[ ]= 3 Cos[th]^2 Sin[th] - Sin[th]^3
```

Expressing z as $re^{i\theta}$ is called polar form while expressing z as $x + yi$ is called rectangular form. The quantities, r and θ, in polar form are called the modulus and the argument of z, respectively.

On the complex plane, each complex number can be visualized as a vector with r as the length and θ as the angle. Thus, the addition and subtraction among complex numbers can be visualized as the addition and subtraction of the corresponding vectors as in Fig. 1.2.

The multiplication and division of two complex numbers can be visualized as in Fig. 1.3. In polar form, z_1 and z_2 can take the following form:

$$z_1 = r_1 e^{i\theta_1}, \quad z_2 = r_2 e^{i\theta_2}. \qquad (1.37)$$

Therefore, the product and the division of z_1 and z_2 are

$$z_1 z_2 = r_1 r_2 e^{i(\theta_1 + \theta_2)}, \quad \frac{z_1}{z_2} = \frac{r_1}{r_2} e^{i(\theta_1 - \theta_2)}. \qquad (1.38)$$

Fig. 1.2 Addition and subtraction of complex numbers

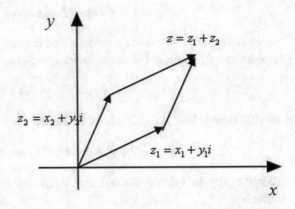

Fig. 1.3 Multiplication of complex numbers

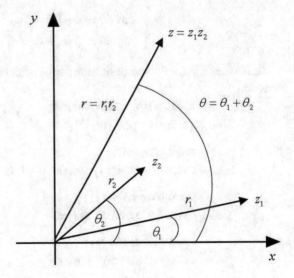

The module of $z = z_1 z_2$ is $r_1 r_2$ and the argument of z is $\theta = \theta_1 + \theta_2$. Similarly, in the division, the module is z_1/z_2 and the argument is $\theta = \theta_1 - \theta_2$ as

$$\arg(z_1 z_2) = \arg z_1 + \arg z_2, \quad \arg \frac{z_1}{z_2} = \arg z_1 - \arg z_2. \tag{1.39}$$

1.3   Complex-Valued Functions

In this section, some important elementary functions of a complex variable are introduced. It is of no surprise that all of the complex functions in this section have their counterparts in real variables thanks to analytic continuation discussed in Sect. 4.1.2.

1.3.1   Exponential Function, e^z

Many textbooks define the exponential function, e^z, by its Taylor series as

$$e^z \equiv 1 + z + \frac{z^2}{2!} + \frac{z^3}{3!} + \dots, \tag{1.40}$$

which is identical to the Taylor series for the real exponential function, e^x, by replacing x by z. We have to wait until Sect. 4.1 when the Taylor/Laurent series is discussed to define e^z this way.

For now, we accept Euler's formula (Eq. (1.25)) and write e^z where $z = x + yi$ as

$$
\begin{aligned}
e^z &= e^{x+yi} \\
&= e^x e^{yi} \\
&= e^x \left(\cos y + i \sin y \right).
\end{aligned}
\tag{1.41}
$$

Thus, the real part of e^z is $e^x \cos y$ and the imaginary part of e^z is $e^x \sin y$.

Similar to e^x, the complex e^z has the following properties:

1.

$$e^{z_1} e^{z_2} = e^{z_1 + z_2}. \tag{1.42}$$

This is known as the law of exponents.

Proof For $z_1 = x_1 + y_1 i$ and $z_2 = x_2 + y_2 i$,

$$e^{z_1} e^{z_2} = \left(e^{x_1} \cos y_1 + i e^{x_1} \sin y_1 \right) \left(e^{x_2} \cos y_2 + i e^{x_2} \sin y_2 \right), \tag{1.43}$$

and

$$e^{z_1 + z_2} = e^{x_1 + x_2} \cos (y_1 + y_2) + i e^{x_1 + x_2} \sin (y_1 + y_2). \tag{1.44}$$

The right-hand sides of Eqs. (1.43) and (1.44) are the same once fully expanded using the trigonometric addition formulas.[3]   □

2.

$$\frac{d}{dz} e^z = e^z. \tag{1.46}$$

This relationship is identical to the corresponding e^x in real variables, i.e., differentiation of the complex exponential function remains the same. Again, the proof and relating topics need to wait until differentiation of complex functions is discussed in Sect. 2.1.

---

[3]

$$\sin (\alpha + \beta) = \sin \alpha \cos \beta + \cos \alpha \sin \beta. \tag{1.45}$$

1.3.2   Trigonometric Functions

In Euler's formula (Eq. (1.25)), the angle, θ, is a real variable. However, if θ in Eq. (1.25) is replaced by a complex variable, z, Eq. (1.25) can be rewritten as

$$e^{iz} = \cos z + i \sin z. \tag{1.47}$$

By replacing z by $-z$, we have

$$e^{-iz} = \cos z - i \sin z. \tag{1.48}$$

From Eqs. (1.47)–(1.48), $\cos z$ and $\sin z$ can be solved as

$$\cos z = \frac{e^{iz} + e^{-iz}}{2}, \tag{1.49}$$

$$\sin z = \frac{e^{iz} - e^{-iz}}{2i}. \tag{1.50}$$

Equations (1.49)–(1.50) can be used to define $\cos z$ and $\sin z$ when z is a complex variable.

Similarly, the hyperbolic sine and cosine functions of a complex variable can be defined as

$$\cosh z = \frac{e^z + e^{-z}}{2}, \tag{1.51}$$

$$\sinh z = \frac{e^z - e^{-z}}{2}. \tag{1.52}$$

The following similarity between trigonometric functions and hyperbolic functions can be immediately verified:

$(\sin z)' = \cos z$	$(\sinh z)' = \cosh z$
$(\cos z)' = -\sin z$	$(\cosh z)' = \sinh z$
$\cos^2 z + \sin^2 z = 1$	$\cosh^2 z - \sinh^2 z = 1$

Example 1

Evaluate

$$z = \sin i. \tag{1.53}$$

What is meant by *evaluate* is to identify the real and imaginary parts of z. Substituting i for z in Eq. (1.50), we have

$$
\begin{aligned}
\sin i &= \frac{e^{ii} - e^{-ii}}{2i} \\
&= \frac{e^{-1} - e}{2i} \\
&= \frac{i}{2}(e - \frac{1}{e}) \\
&\sim 1.1752\,i.
\end{aligned}
\tag{1.54}
$$

Note that this is a pure imaginary number.

Example 2

Evaluate

$$
I \equiv \int_0^\infty e^{-x} \cos x dx.
\tag{1.55}
$$

If we are not allowed to use complex variables, it is necessary to employ integration by parts twice to carry out I. However, the use of Euler's formula can avoid the integration by parts altogether. Let a companion integral to I, J, be defined as

$$
J \equiv \int_0^\infty e^{-x} \sin x dx.
\tag{1.56}
$$

It follows

$$
\begin{aligned}
I + iJ &= \int_0^\infty e^{-x} \left(\cos x + i \sin x \right) dx \\
&= \int_0^\infty e^{-x} e^{ix} dx \\
&= \int_0^\infty e^{(-1+i)x} dx \\
&= \left[\frac{1}{-1+i} e^{(-1+i)x} \right]_0^\infty \\
&= \left(\frac{1}{-1+i}(0 - 1) \right) \\
&= \left(\frac{-1}{-1+i} \frac{-1-i}{-1-i} \right) \\
&= \frac{1}{2} + \frac{1}{2}i.
\end{aligned}
\tag{1.57}
$$

As I is the real part[4] of Eq. (1.57), it follows

$$I = \frac{1}{2}. \tag{1.58}$$

As a bonus, J is also evaluated as

$$J = \frac{1}{2}. \tag{1.59}$$

Example 3

Solve

$$\sin z = 3, \tag{1.60}$$

for z.

This equation does not make sense if z is a real number as $\sin x$ is bounded between -1 and 1. However, complex trigonometric functions have no such restriction. Let $z = x + yi$, then we have

$$\begin{aligned}
\sin z &= \sin(x + yi) \\
&= \frac{1}{2i}\left(e^{i(x+yi)} - e^{-i(x+yi)}\right) \\
&= \frac{1}{2i}\left(\cos x(e^{-y} - e^{y}) + i\sin x(e^{-y} + e^{y})\right).
\end{aligned} \tag{1.61}$$

Therefore, $\sin z = 3$ is equivalent to

$$\cos x(e^{-y} - e^{y}) + i\sin x(e^{-y} + e^{y}) = 6i. \tag{1.62}$$

Comparing the imaginary part and real part of both sides yields

$$\cos x(e^{-y} - e^{y}) = 0, \tag{1.63}$$
$$\sin x(e^{-y} + e^{y}) = 6. \tag{1.64}$$

If $e^{-y} - e^{y} = 0$ in Eq. (1.63), it follows $e^{2y} = 1$ or $y = 0$ which implies that $\sin x = 3$ from Eq. (1.64). As this is not possible, $\cos x = 0$ must be satisfied from Eq. (1.63). Therefore,

$$x = \frac{\pi}{2} \pm n\pi, \tag{1.65}$$

where n is an integer and

$$\sin\left(\frac{\pi}{2} \pm n\pi\right)(e^{-y} + e^{y}) = 6. \tag{1.66}$$

For the left-hand side of Eq. (1.66) to be positive, n must be an even integer as

$$x = \frac{\pi}{2} \pm 2n\pi. \tag{1.67}$$

---

[4] As $x \to \infty$, $e^{(-1+i)x} = e^{-x}e^{ix} \to 0$ since $e^{-x} \to 0$ and $e^{ix} = \cos x + i\sin x$ is bounded (but indefinite).

From Eq. (1.66), it follows

$$e^{-y} + e^y = 6,\qquad(1.68)$$

or

$$(e^y)^2 - 6e^y + 1 = 0,\qquad(1.69)$$

which can be solved for y as

$$y = \log(3 \pm 2\sqrt{2}).\qquad(1.70)$$

Finally, we have

$$z = \left(\frac{\pi}{2} \pm 2n\pi\right) + i\,\log(3 \pm 2\sqrt{2}).\qquad(1.71)$$

1.3.3 Logarithmic Function, $\log z$

In real-valued functions, the real logarithmic function, $\log x$, is defined as the inverse function of e^x, i.e., $y = \log x$ and $x = e^y$ are equivalent. As e^y is always positive definite, the argument of log must be also positive definite. The logarithmic function of a complex variable, $\log z$, is a complex number, and most importantly, it is multi-valued.

We start with expressing a complex number, z, in polar form as

$$z = r\,e^{i(\theta+2n\pi)},\qquad(1.72)$$

where n is an integer. Although adding the extra $2n\pi$ to θ causes no change on the value of z, this makes $\log z$ multi-valued.

Assuming the real logarithmic rule applies, taking log on Eq. (1.72) yields

$$\begin{aligned}
\log z &= \log\left(re^{i(\theta+2n\pi)}\right) \\
&= \operatorname{Log} r + \log e^{i(\theta+2n\pi)} \\
&= \operatorname{Log} r + i(\theta + 2n\pi)\log_e e \\
&= \operatorname{Log} r + i(\theta + 2n\pi).
\end{aligned}\qquad(1.73)$$

To avoid confusion, we use log for the logarithm of a complex number and Log for the natural logarithm of a real (and positive) number.

For z given as Eq. (1.72), the real part of $\log z$ is $\operatorname{Log} r$ and the imaginary part is $\theta + 2n\pi$. Unless the value of n is specified, $\log z$ cannot be uniquely determined. For instance, $\operatorname{Log} 1 = 0$. However, $\log 1 = 2n\pi i$ as $1 = 1e^{2n\pi i}$ in polar form. It is necessary to know which logarithm is to be used. When $n = 0$, $\log z$ is called the principal value.

Example 1
Evaluate

$$z = \log i.\qquad(1.74)$$

Since i on the complex plane is identified as a vector with the angle, $\frac{\pi}{2} + 2n\pi$, and the magnitude of 1 as

$$i = 1e^{i(\pi/2+2n\pi)}, \tag{1.75}$$

it follows

$$\begin{aligned}
\log i &= \log \left(1e^{i(\pi/2+2n\pi)}\right) \\
&= \mathrm{Log}\, 1 + i\left(\frac{\pi}{2} + 2n\pi\right) \log_e e \\
&= \mathrm{Log}\, 1 + i\left(\frac{\pi}{2} + 2n\pi\right) \\
&= i\left(\frac{\pi}{2} + 2n\pi\right).
\end{aligned} \tag{1.76}$$

Note that this is a pure imaginary number and multi-valued.

Example 2

Evaluate

$$z = 1^i. \tag{1.77}$$

By taking logarithm on both sides, we have

$$\begin{aligned}
\log z &= \log 1^i \\
&= i \log 1 \\
&= i\, (i\,(2n\pi)) \\
&= 2\, m\pi,
\end{aligned} \tag{1.78}$$

where $m(= -n)$ is an integer. Therefore,

$$1^i = e^{2\,m\pi}. \tag{1.79}$$

Note that 1^i is a real (positive) number.

Example 3

Evaluate

$$z = (1 + i)^i. \tag{1.80}$$

By taking logarithm on both sides, we have

$$\begin{aligned}
\log z &= \log (1 + i)^i \\
&= i \log (1 + i) \\
&= i \log \left(\sqrt{2}e^{(i\frac{\pi}{4}+2n\pi i)}\right) \\
&= i\left(\mathrm{Log}\, \sqrt{2} + i\left(\frac{\pi}{4} + 2n\pi\right)\right),
\end{aligned} \tag{1.81}$$

where we used

$$1 + i = \sqrt{2} \left(\frac{1}{\sqrt{2}} + i \frac{1}{\sqrt{2}} \right)$$
$$= \sqrt{2} \left(\cos \frac{\pi}{4} + i \sin \frac{\pi}{4} \right)$$
$$= \sqrt{2} e^{(\frac{\pi}{4} + 2n\pi)i}. \tag{1.82}$$

Therefore,

$$z = e^{-(\frac{\pi}{4} + 2n\pi)} e^{i \operatorname{Log} \sqrt{2}}$$
$$= e^{-(\frac{\pi}{4} + 2n\pi)} \left(\cos (\operatorname{Log}\sqrt{2}) + i \sin (\operatorname{Log} \sqrt{2}) \right). \tag{1.83}$$

1.3.4 Branch Cut and Branch Points

As was the case with $\log z$, functions of a complex variable can be multi-valued. Other examples of multi-valued functions include z^{α} where α is a non-integer. For example, consider

$$f(z) = z^{\frac{1}{2}} (= \sqrt{z}). \tag{1.84}$$

By expressing z in polar form, we have

$$\sqrt{z} = \left(r e^{i(\theta + 2n\pi)} \right)^{\frac{1}{2}}$$
$$= \sqrt{r} \, e^{i(\theta + 2n\pi)/2}$$
$$= \sqrt{r} \, e^{i\theta/2} e^{n\pi i}, \tag{1.85}$$

where $\sqrt{r}$ is the square root of a real positive number. The term, $e^{n\pi i}$, is -1 if n is an odd number and 1 if n is an even number. Therefore,

$$\sqrt{z} = \begin{cases} \sqrt{r} \, e^{i\theta/2}, & \text{if } n \text{ is even,} \\ -\sqrt{r} \, e^{i\theta/2}, & \text{if } n \text{ is odd.} \end{cases} \tag{1.86}$$

For instance, $\sqrt{2} = 1.41421\ldots$ if we express "2" as $2e^{4\pi i}$ but $\sqrt{2} = -1.41421\ldots$ if we express "2" as $2e^{2\pi i}$.

This ambiguity is caused by n, the number of rotations about $z = 0$, in the argument of polar form, $\theta + 2n\pi$. If we can impose a condition that n must be 0, $\sqrt{z}$ is uniquely evaluated. For this purpose, branch cuts and branch points are introduced to avoid multi-valuedness of complex functions.

If we make a cut on the real axis as shown in Fig. 1.4 starting at $z = 0$ extending to $z \to \infty$ so that a complex number, z, cannot travel across the cut, the range of the argument of z is limited as $0 < \theta < 2\pi$.

The line segment from $z = 0$ to $z \to \infty$ is called a branch cut and the starting point, $z = 0$, is called a branch point.

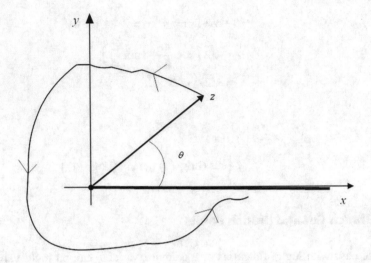

Fig. 1.4 A branch cut for $\sqrt{z}$

This selection of the branch cut and the branch point is to effectively force $n = 0$. It is noted that branch cuts and branch points depend on $f(z)$ and are not unique. Any one of the three curves in Fig. 1.5 is a valid branch cut and branch point.

However, branch cuts and branch points shown in Fig. 1.6 do not work for $\sqrt{z}$ as the vector can rotate around the origin as many times avoiding the cuts.

From the examples above, the conditions of a proper branch cut for $\sqrt{z}$ are as follows:

1. The branch cut must pass the origin.
2. The branch cut must be extended to infinity.

As a slight variation to $\sqrt{z}$, a branch cut and branch point for $f(z) = \sqrt{z-a}$ are shown in Fig. 1.7.

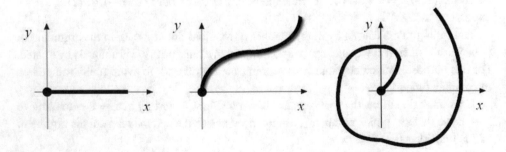

Fig. 1.5 Possible branch cuts for $\sqrt{z}$

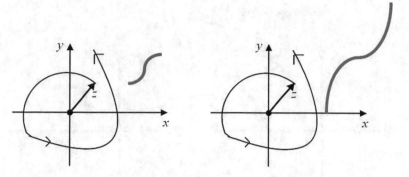

Fig. 1.6 Branch cuts that do not work for $\sqrt{z}$

Fig. 1.7 A branch cut for
$\sqrt{z-a}$

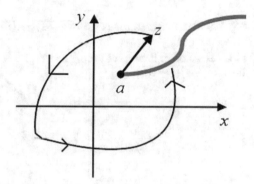

As an another example of branch cut/point, consider

$$f(z) = \sqrt{(z-a)(z-b)}. \tag{1.87}$$

A branch cut for $f(z)$ can be a combination of the branch cuts for $\sqrt{z-a}$ and $\sqrt{z-b}$ as shown in the left figure of Fig. 1.8.

The right figure of Fig. 1.8 is another possible branch cut and is a finite-length curve between a and b. In this case, a branch cut does not have to be extended to infinity. We can show that this is a valid branch cut by first expressing $z-a$ and $z-b$ in polar form as

$$z - a = r_1 e^{i\theta_1}, \tag{1.88}$$
$$z - b = r_2 e^{i\theta_2}, \tag{1.89}$$
$$\sqrt{(z-a)(z-b)} = (r_1 r_2)^{\frac{1}{2}} e^{i\frac{(\theta_1+\theta_2)}{2}}. \tag{1.90}$$

From Eq. (1.90), the argument of $\sqrt{(z-a)(z-b)}$ is $(\theta_1 + \theta_2)/2$. As shown in Fig. 1.9, when both θ_1 and θ_2 are incremented by 2π after one full rotation around the finite-length branch cut, the increment of $(\theta_1 + \theta_2)/2$ is also 2π, which does not affect the value of $e^{i(\theta_1+\theta_2)/2}$.

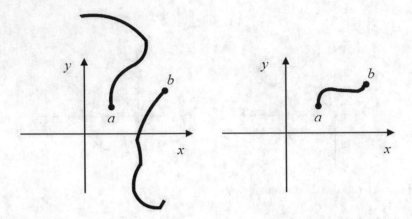

Fig. 1.8 Possible branch cuts for $\sqrt{(z-a)(z-b)}$

Fig. 1.9 Another branch cut
for $\sqrt{(z-a)(z-b)}$

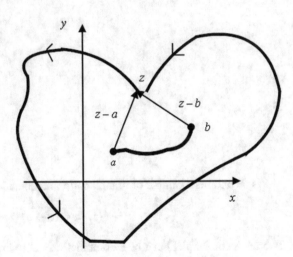

Summary

Branch cuts are required for some complex functions if they are variations of z^α where α is a non-integer or $\log z$. Branch cuts are not required for z^n where n is an integer. However, exceptions for the non-integer rule of α do exist. For instance, consider $f(z) = \cos \sqrt{z}$. It appears that $f(z)$ requires a branch cut because of $\sqrt{z}$. However, the Taylor series of $\cos \sqrt{z}$ is

$$\cos \sqrt{z} = 1 - \frac{z}{2!} + \frac{z^2}{4!} - \frac{z^4}{6!} + \cdots . \tag{1.91}$$

It is seen that the terms on the right-hand side of Eq. (1.91) do not require branch cuts. Hence, $\cos \sqrt{z}$ does not require branch cuts. Note that $\sin \sqrt{z}$ does require a branch cut as its Taylor series contains $\sqrt{z}$.

1.4 Numerics

Mathematica can be used to visualize complex numbers [7].

In[ ]:= **Sin[2 + 3 I]**

Out[ ]= Sin[2 + 3 i]

Entering `Sin[2+3I]` returns itself unevaluated. If you want to see the real part and the imaginary part, use the `ComplexExpand[]` function as

In[ ]:= **ComplexExpand[Sin[2 + 3 I]]**

Out[ ]= Cosh[3] Sin[2] + i Cos[2] Sinh[3]

To see the numerical value (floating number), use the `N[]` function that converts the symbolic value to the numerical value.

In[ ]:= **N[Sin[2 + 3 I]]**

Out[ ]= 9.1545 – 4.16891 i

Use `Abs[]` to see the absolute value of a complex number as

In[ ]:= **Abs[%]**

Out[ ]= 10.0591

where the percentage sign (%) represents the preceding output.

The absolute value of $\sin(x + yi)$ can be visualized by the following command:

In[ ]:= **Plot3D[Abs[Sin[x + I y]], {x, -10, 10}, {y, -10, 10}]**

The `Plot3D[f, x, xmin, xmax, y, ymin, ymax]` function draws a 3-D plot for $f(x, y)$ over $(xmin < x < xmax)$ and $(ymin < y < ymax)$. It is noted that $|\sin z|$ keeps increasing beyond 1 on the y-axis.

If we nest the sine function, its values increase quite rapidly. To see its magnitude, we can take the natural logarithm twice on the result of the nested sins.

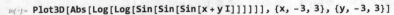

In[ ]:= **Plot3D[Abs[Log[Log[Sin[Sin[Sin[x + y I]]]]]], {x, -3, 3}, {y, -3, 3}]**

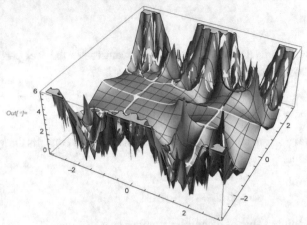

It is seen that bifurcating patterns are formed along the edge on the y-axis which can be seen more clearly if we use the `DensityPlot[]` function.

In[ ]:= **DensityPlot[Abs[Log[Log[Sin[Sin[Sin[x + y I]]]]]], {x, -3, 3}, {y, -3, 3}]**

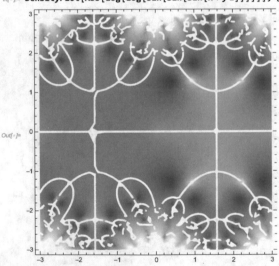

Reference [7] has interesting discussions on this topic.

1.5 Problems

1. Solve
$$\sin z = -2. \tag{1.92}$$

2. Evaluate
$$\left(\frac{1+i}{1-i}\right)^{100}. \tag{1.93}$$

3. Evaluate
$$\cos(1+i). \tag{1.94}$$

4. Find the real and imaginary parts for
$$i^{i+1}. \tag{1.95}$$

5. Evaluate
$$\int_0^\infty e^{-ax} \sin bx\, dx, \quad a > 0 \tag{1.96}$$

   using complex variables.

6. Show

   (a)
   $$|\sin z|^2 = \frac{1}{2}(\cosh 2y - \cos 2x). \tag{1.97}$$

   (b)
   $$|\cos z|^2 = \sin^2 x \, \sinh^2 y + \cos^2 x \, \cosh^2 y. \tag{1.98}$$

Calculus of Functions of Complex Variables

2

2.1 Differentiability (Cauchy-Riemann Equations)

A function of a real variable, $f(x)$, can be differentiable at a point $x = x_0$ if the following limit exists and is unique:

$$\lim_{h \to 0} \frac{f(x_0 + h) - f(x)}{h}, \tag{2.1}$$

where h approaches to 0 independent of whether it is coming from $+\infty$ or $-\infty$. The step function defined as

$$f(x) = |x| \tag{2.2}$$

is not differentiable at $x = 0$ because

$$\lim_{h \to 0} \frac{f(0 + h) - f(0)}{h} = \frac{|h|}{h} = \begin{cases} 1, & h > 0 \\ -1, & h < 0, \end{cases} \tag{2.3}$$

i.e., if h approaches to 0 from $+\infty$, the quotient is 1 and if h approaches to 0 from $-\infty$, the quotient is -1. They do not match at $h = 0$. This is also obvious from the graph of $f(x)$ as the slopes of $f(x)$ do not match at $x = 0$.

Differentiability of a function of a complex variable, $f(z)$, can be defined similar to Eq. (2.1). A complex function, $f(z)$, is differentiable at $z = z_o$ when the following differential quotient has a limit and converges to a unique value as

$$\lim_{\Delta z \to 0} \frac{f(z_0 + \Delta z) - f(z_0)}{\Delta z} = f'(z_0). \tag{2.4}$$

The difference between Eqs. (2.1) and (2.4) is that in Eq. (2.1), h is restricted on the x-axis and can approach to 0 from either $+\infty$ or $-\infty$ while in Eq. (2.4), there are infinite ways that Δz can approach to 0 on the complex plane. The condition that the limit of Eq. (2.4) exists

© The Author(s), under exclusive license to Springer Nature Switzerland AG 2022
S. Nomura, *Complex Variables for Engineers with Mathematica*,
Synthesis Lectures on Mechanical Engineering,
https://doi.org/10.1007/978-3-031-13067-0_2

and is unique is so restrictive that only a handful of complex functions are differentiable.
Equation (2.4) must be held regardless of how Δz goes to 0.

Consider a complex function, $f(z)$, defined as

$$f(z) \equiv \bar{z} = x - iy. \tag{2.5}$$

We examine whether $f(z)$ is differentiable or not at $z_0 = 0$. By choosing $z_0 = 0$ in Eq. (2.4),
we have

$$\begin{aligned} f'(0) &= \lim_{\Delta z \to 0} \frac{f(0 + \Delta z) - f(0)}{\Delta z} \\ &= \lim_{\Delta z \to 0} \frac{\overline{\Delta z}}{\Delta z} \\ &= \lim_{\Delta x, \Delta y \to 0} \frac{\Delta x - i\Delta y}{\Delta x + i\Delta y}. \end{aligned} \tag{2.6}$$

If $\Delta z \to 0$ from $+\infty$ on the real axis, $\Delta z = \Delta x$ and $\Delta y = 0$. Therefore, Eq. (2.6) becomes

$$\lim_{\Delta x \to 0} \frac{\Delta x}{\Delta x} = 1. \tag{2.7}$$

On the other hand, if $\Delta z \to 0$ from $+\infty$ on the imaginary axis, $\Delta z = i\Delta y$ and $\Delta x = 0$.
Therefore, Eq. (2.6) becomes

$$\lim_{\Delta y \to 0} \frac{-i\Delta y}{i\Delta y} = -1. \tag{2.8}$$

If Δz approaches to 0 along $y = x$, $\Delta x = \Delta y$. Hence, Eq. (2.6) becomes

$$\lim_{\Delta x \to 0} \frac{\Delta x(1 - i)}{\Delta x(1 + i)} = \frac{1 - i}{1 + i}. \tag{2.9}$$

As is seen from Eqs. (2.7)–(2.9), $f(z) \equiv x - iy$ is not differentiable at $z = 0$ as Eqs. (2.7)–
(2.9) do not match. Even though both the real part and the imaginary part of $f(z)$ are perfectly
differentiable as functions of x and y, their combination is not necessarily so. There must
be a relationship that the real and imaginary parts of a complex function must satisfy. Such
a relationship is known as the Cauchy-Riemann equations.

2.1.1 Cauchy-Riemann Equations

For a complex function, $f(z) = u + iv$, to be differentiable at $z = z_0$, the real part of $f(z)$,
$u(x, y)$, and the imaginary part, $v(x, y)$, must satisfy the following equations at $z = z_0$
called the Cauchy-Riemann equations:

$$\frac{\partial u}{\partial x} = \frac{\partial v}{\partial y}, \quad \frac{\partial u}{\partial y} = -\frac{\partial v}{\partial x}. \tag{2.10}$$

Proof For $f(z) = u + iv$ to be differentiable at $z = z_0$, the limit of the differential quotient, $\frac{\Delta f}{\Delta z}$, must exist and be unique as $\Delta z \to 0$ regardless of how Δz approaches to z_0. Thus, the limit is expressed as

$$\lim_{z \to z_0} \frac{\Delta f}{\Delta z} = A + B i, \tag{2.11}$$

where A and B are real numbers.

Equation (2.11) can be written as

$$\Delta f \sim (A + B i)\Delta z, \tag{2.12}$$

or

$$\left(u_x \Delta x + u_y \Delta y\right) + i \left(v_x \Delta x + v_y \Delta y\right) \sim (A + B i)(\Delta x + i\Delta y), \tag{2.13}$$

where $u_x = \frac{\partial u}{\partial x}$, etc. If we compare the real and imaginary parts of both sides of Eq. (2.13), the following relations are derived:

$$u_x \Delta x + u_y \Delta y \sim A\Delta x - B\Delta y, \tag{2.14}$$

$$v_x \Delta x + v_y \Delta y \sim B\Delta x + A\Delta y. \tag{2.15}$$

By comparing the coefficients of Δx and Δy above, we have

$$u_x = A, \quad u_y = -B, \tag{2.16}$$

$$v_x = B, \quad v_y = A, \tag{2.17}$$

which are equivalent to

$$u_x = v_y, \quad u_y = -v_x. \tag{2.18}$$

$$\square$$

Equations (2.18) are the Cauchy-Riemann equations.

If $f(z)$ is differentiable everywhere in a domain D, we say that $f(z)$ is analytic in that domain. For instance, consider a function, $f(z) \equiv |z|^2 = x^2 + y^2$. It follows

$$u = x^2 + y^2, \quad v = 0. \tag{2.19}$$

Applying the Cauchy-Riemann equations on Eq. (2.19) yields

$$2x = 0, \quad 2y = 0, \tag{2.20}$$

which are satisfied only when $x = y = 0$. Therefore, $f(z)$ is differentiable at $z = 0$ but not analytic in the neighborhood of $z = 0$.

There are other terminologies for $f(z)$ besides *analytic*. Some textbooks use *holomorphic* and other use *regular*. We will use *analytic* in this book.

If $f(z)$ is not analytic at z_0, we say that z_0 is a singular point or singularity.

The Cauchy-Riemann equations of Eq. (2.18) imply that if $f(z)$ is analytic, its real part, u, and imaginary part, v, are not independent of each other. You cannot make an analytic function, $f(z)$, by simply combining two real-valued and differentiable functions, $u(x, y)$ and $v(x, y)$, together as $f = u + iv$. For instance, even though $u = e^x \sin y$ and $v = x^2 y - y^3 \cos xy$ are both differentiable as functions of x and y, their combination of $f = (e^x \sin y) + i(x^2 y - y^3 \cos xy)$ is not analytic, which can be easily checked by testing the Cauchy-Riemann equations. In fact, as will be shown, once $u(x, y)$ of an analytic function is provided, $v(x, y)$ is automatically determined except for a constant. Such restrictions significantly limit the number of available analytic functions which is the main reason why analytic functions play an important role in complex variables.

Examples

1. $f(z) = z^3$.
   As $u = x^3 - 3xy^2$ and $v = 3x^2 y - y^3$, $u_x = v_y$ and $u_y = -v_x$ are satisfied. Hence, $f(z) = z^3$ is analytic.
2. $f(z) = (x^2 - y^3) + 2(e^x - \sin y)i$.
   The Cauchy-Riemann equations, $u_x = v_y$ and $u_y = -v_x$, are not satisfied.
3. $f(z) = 1/z$.
   As $u = \frac{x}{x^2+y^2}$ and $v = -\frac{y}{x^2+y^2}$, $u_x = v_y$ and $u_y = -v_x$ are satisfied. Hence, $f(z) = \frac{1}{z}$ is analytic except for $z = 0$.
4. $f(z) = z^{100}$.
   Expanding $(x + iy)^{100}$, identifying u and v and applying the Cauchy-Riemann equations are not practical. *Mathematica* can expand $(x + iy)^{100}$ analytically without a problem and verify the Cauchy-Riemann equations. However, there exists an alternative way to check the Cauchy-Riemann equations which does not require the separation of $f(z)$ into u and v as shown in Sect. 2.1.2.

Mathematica **Code**

The following *Mathematica* code can check the Cauchy-Riemann equations. A user-defined function such as $f(z) \equiv z^3$ can be entered as f[z_]:=z ^ 3 where the symbol with the underline in the left-hand side is the variable of the function. Note the colon and the equal sign. After this definition, f[z] can be used as if it were a built-in function. Differentiation of $f(x)$ with respect to x can be entered as D[f[x], x].

In[ ]:= **z = x + I y**

Out[ ]= **x + i y**

In[ ]:= **u = z^3 // Re // ComplexExpand**

Out[ ]= **x³ − 3 x y²**

In[ ]:= **v = z^3 // Im // ComplexExpand**

Out[ ]= **3 x² y − y³**

In[ ]:= **D[u, x] == D[v, y]**

Out[ ]= **True**

In[ ]:= **D[u, y] == −D[v, x]**

Out[ ]= **True**

The statement, D[u,x]==D[v,y], in the code above means whether D[u,x] and D[v,y] are the same or not. It returns either True or False. As in any programming language, a single equal sign (=) is an assignment from right to left and double equal signs (==) are an equality.

2.1.2   Alternative Form of Cauchy-Riemann Equations

The Cauchy-Riemann equations are not effective if the real part and the imaginary part of a complex function are not easily separated such as $f(z) = z^{100}$. We can derive an alternative to the Cauchy-Riemann equations which does not require separation of the real and imaginary parts of a complex function.

A complex function, $f = u(x, y) + i\, v(x, y)$, can be seen as a function of x and y through $u(x, y)$ and $v(x, y)$. A pair of (x, y) is interchangeable with $(z, \bar{z})$ as

$$z = x + yi, \quad \bar{z} = x - yi, \tag{2.21}$$

$$x = \frac{1}{2}(z + \bar{z}), \; y = \frac{1}{2i}(z - \bar{z}). \tag{2.22}$$

With Eq. (2.22), z and $\bar{z}$ can be taken as two independent variables instead of x and y and f can be viewed as a function of z and $\bar{z}$. Using the chain differentiation rule, we can differentiate f with respect to $\bar{z}$ implicitly even if f is a function of (x, y) as

$$\begin{aligned}
\frac{\partial f(z, \bar{z})}{\partial \bar{z}} &= \frac{\partial f}{\partial x}\left(\frac{\partial x}{\partial \bar{z}}\right) + \frac{\partial f}{\partial y}\left(\frac{\partial y}{\partial \bar{z}}\right) \\
&= (u_x + i v_x)\frac{1}{2} - (u_y + i v_y)\frac{1}{2i} \\
&= (v_y + i v_x)\frac{1}{2} - (-v_x + i v_y)\frac{1}{2i} \\
&= 0,
\end{aligned} \tag{2.23}$$

where the Cauchy-Riemann equations, $u_x = v_y$, $u_y = -v_x$, were used.

The result of Eq. (2.23) implies that if the Cauchy-Riemann equations are held, f must be independent of $\bar{z}$, i.e., f does not contain $\bar{z}$. This is an alternative form to the Cauchy-Riemann equations. If f is found to contain $\bar{z}$, we can immediately conclude that f is not differentiable.

Examples

1. $f = |z|^2$.
   This is not differentiable as $|z|^2 = z\bar{z}$.
2. $f = z^{100}$.
   This is differentiable as f does not contain $\bar{z}$.
3. $f = \bar{z}$.
   This is not differentiable as f does contain $\bar{z}$.

2.1.3   Harmonic Functions

One of the reasons why complex functions are useful in various problems in engineering is because the real and imaginary parts of an analytic function automatically satisfy the Laplace equation to which many equations in engineering are subject.

A harmonic function (2-D), $f(x, y)$, is a function that satisfies the Laplace equation, i.e.,

$$\Delta f(x, y) \equiv \frac{\partial^2 f(x, y)}{\partial x^2} + \frac{\partial^2 f(x, y)}{\partial y^2}$$

$$= 0. \tag{2.24}$$

If $f(z) = u + iv$ is analytic, its real part, $u(x, y)$, and imaginary part, $v(x, y)$, both satisfy the Laplace equation in 2-D, i.e., both are harmonic functions.

Proof By differentiating both sides of $u_x = v_y$ (the first of the C-R equations) with respect to x, we have $u_{xx} = v_{yx}$, and by differentiating both sides of $u_y = -v_x$ (the second of the C-R equations) with respect to y, we have $u_{yy} = -v_{xy}$. Adding the two yields $u_{xx} + u_{yy} = 0$. Similar for $v_{xx} + v_{yy} = 0$.   □

This result implies that we can easily find a harmonic function by simply selecting an arbitrary analytic function and by extracting its real and imaginary parts. Of course, we must also consider the boundary condition of the concerned boundary value problem as a harmonic function alone does not satisfy the prescribed boundary condition. One of the techniques for the solution to a boundary value problem where the governing equation is the Laplace equation (or its variations) is to express the solution by a combination of analytic functions in power series so that the combination of analytic functions satisfies the boundary conditions. This technique will be discussed in Sect. 6.2.

Example 1

$$f(z) \equiv z^3$$
$$= (x + iy)^3$$
$$= (x^3 - 3xy^2) + (3x^2y - y^3)i. \tag{2.25}$$

The real and imaginary parts of $f(z)$ are

$$u = x^3 - 3xy^2,$$
$$v = 3x^2y - y^3. \tag{2.26}$$

It is easy to verify that $\Delta u = \Delta v = 0$.

Example 2

$$f(z) \equiv \frac{1}{z}$$
$$= \frac{\bar{z}}{z\bar{z}}$$
$$= \frac{x - iy}{x^2 + y^2}. \tag{2.27}$$

The real and imaginary parts of $f(z)$ are

$$u = \frac{x}{x^2 + y^2},$$
$$v = \frac{-y}{x^2 + y^2}. \tag{2.28}$$

It is easy to verify that $\Delta u = \Delta v = 0$.

Mathematica **Code**

The following *Mathematica* code implements the above:

```
In[*]:= z = x + I y;

In[*]:= realpart = ComplexExpand[Re[z^3]]

Out[*]= x³ - 3 x y²

In[*]:= impart = ComplexExpand[Im[z^3]]

Out[*]= 3 x² y - y³

In[*]:= laplace[f_] := D[f, {x, 2}] + D[f, {y, 2}]

In[*]:= laplace[realpart]

Out[*]= 0

In[*]:= laplace[impart]

Out[*]= 0
```

2.1.4   Uniqueness of Analytic Functions

The real and imaginary parts of an analytic function are not Independent of each other as they must satisfy the Cauchy-Riemann equations. Moreover, once the real part is provided, the imaginary part is automatically determined and vice versa except for an addition of a constant.

Example 1
Given $u(x, y) = x^3 - 3xy^2$, find $v(x, y)$.

From the Cauchy-Riemann equations, $u_x = v_y$ and $u_y = -v_x$, it follows

$$v_y = 3x^2 - 3y^2, \tag{2.29}$$

$$v_x = -(-6xy). \tag{2.30}$$

By integrating Eq. (2.30) with respect to x, we have

$$v = 3x^2y + f(y), \tag{2.31}$$

where $f(y)$ is a function of y alone. By substituting Eq. (2.31) into Eq. (2.29), we have

$$3x^2 + f'(y) = 3x^2 - 3y^2, \tag{2.32}$$

i.e.,

$$f(y) = -y^3 + C, \tag{2.33}$$

where C is a constant. Thereby,

$$v = 3x^2y - y^3 + C. \tag{2.34}$$

It is seen that

$$\begin{aligned} u + iv &= (x^3 - 3xy^2) + i(3x^2y - y^3) + C \\ &= (x + iy)^3 + C \\ &= z^3 + C, \end{aligned}$$ (2.35)

which is analytic (no $\bar{z}$).

Example 2
Given

$$u = e^{2x} \cos 2y,$$ (2.36)

find $v(x, y)$.

Without the Cauchy-Riemann equations, we can use an educated guess by choosing $v = e^{2x} \sin 2y$ as a companion to u. To verify whether this is a correct guess, combine u and v through i as

$$\begin{aligned} u + iv &= e^{2x}(\cos 2y + i \sin 2y) \\ &= e^{2x} e^{2yi} \\ &= e^{2(x+iy)} \\ &= e^{2z}, \end{aligned}$$ (2.37)

which is analytic (no $\bar{z}$), revealing that our guess was correct.

Example 3
Given

$$u = e^{2x} \sin 2y,$$ (2.38)

find $v(x, y)$.

From Example 2, we may be tempted to choose

$$v = e^{2x} \cos 2y.$$ (2.39)

However, this does not work as

$$u + iv = e^{2x} \sin 2y + ie^{2x} \cos 2y,$$ (2.40)

which cannot be reduced to any simpler form that has the factor of $\cos 2y + i \sin 2y$. Nevertheless, by slightly modifying the selection, we can try

$$v = -e^{2x} \cos 2y.$$ (2.41)

We have

$$\begin{aligned}
u + iv &= e^{2x}(\sin 2y - i \cos 2y) \\
&= e^{2x}(-i^2 \sin 2y - i \cos 2y) \\
&= e^{2x}(-i)(\cos 2y + i \sin 2y) \\
&= -ie^{2x}e^{2yi} \\
&= -ie^{2(x+iy)} \\
&= -ie^{2z},
\end{aligned} \tag{2.42}$$

which is a function of z alone.

Example 4

Given

$$u = r^2 \cos 2\theta + 4, \tag{2.43}$$

find $v(x, y)$.

By inspection, we can try

$$v = r^2 \sin 2\theta, \tag{2.44}$$

so that

$$\begin{aligned}
u + iv &= r^2(\cos 2\theta + i \sin 2\theta) + 4 \\
&= r^2 e^{2\theta i} + 4 \\
&= z^2 + 4,
\end{aligned} \tag{2.45}$$

which is a function of z alone.

2.2   Problems

1. The real part of an analytic function, $f(z)$, is given below. Find the imaginary part, $v(x, y)$, and $f(z)$ as a function of z.

(a)

$$u = \frac{x^2 - 2x - y^2 + 1}{\left(x^2 - 2x + y^2 + 1\right)^2}. \tag{2.46}$$

(b)

$$u = \frac{x^2 - y^2}{(x^2 + y^2)^2}. \tag{2.47}$$

(c)

$$u = e^x(x \cos y - y \sin y). \tag{2.48}$$

(d)

$$u = e^{x^2 - y^2}(x \cos(2xy) - y \sin(2xy)). \tag{2.49}$$

2. Derive the Cauchy-Riemann equations in the polar coordinate system (r, θ), i.e.,

$$\frac{\partial u}{\partial r} = \frac{1}{r}\frac{\partial v}{\partial \theta}, \quad \frac{\partial u}{\partial \theta} = -r\frac{\partial v}{\partial r}. \tag{2.50}$$

3. Show that if $|f(z)| = const$, $f(z)$ must be a constant.

Integrations of Functions of Complex Variables

3

3.1 Integral Calculus

A complex integral is an integral of $f(z)$ from A to B along the prescribed path on the complex plane as shown in Fig. 3.1 and can be expressed as a sum of two line integrals as

$$\int_A^B f(z)dz = \int_A^B (u + iv)(dx + idy)$$

$$= \int_A^B (udx - vdy) + i \int_A^B (udy + vdx). \tag{3.1}$$

The value of a complex integral depends not only on the start and end points but also on the integral path. In general, the values of complex integrals along two different paths are different even though the start and the end are the same.

Example 1

Integrate $f(z) = |z|^2$ from $z = 0$ to $z = 1 + i$ along the two different paths shown in Fig. 3.2.

(1) Along $y = x$ (shortest cut).

Along $y = x$, $dy = dx$. Therefore, y can be replaced by x and dy can be replaced by dx as

$$\int |z|^2 dz = \int (x^2 + y^2)(dx + idy)$$

$$= \int_0^1 (x^2 + x^2)(1 + i)dx$$

$$= 2(1 + i) \int_0^1 x^2 dx$$

$$= \frac{2(1 + i)}{3}. \tag{3.2}$$

© The Author(s), under exclusive license to Springer Nature Switzerland AG 2022
S. Nomura, *Complex Variables for Engineers with Mathematica*,
Synthesis Lectures on Mechanical Engineering,
https://doi.org/10.1007/978-3-031-13067-0_3

Fig. 3.1 Complex integral as a line integral

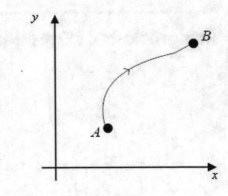

Fig. 3.2 Integration of $|z|^2$ along two different paths

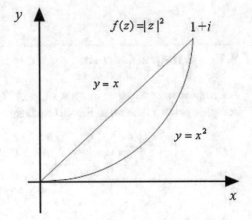

(2) Along $y = x^2$.

Along $y = x^2$, $dy = 2xdx$. Therefore, y can be replaced by x^2 and dy can be replaced by $2x\,dx$ as

$$\int |z|^2 dz = \int (x^2 + y^2)(dx + idy)$$

$$= \int_0^1 (x^2 + x^4)(dx + i2xdx)$$

$$= \int_0^1 (x^2 + x^4)(1 + 2ix)dx$$

$$= \frac{8}{15} + \frac{5}{6}i. \tag{3.3}$$

As the two paths are different, their values are also different as expected. We note that $|z|^2 = z\bar{z}$ is not analytic.

Example 2

Integrate $f(z) = z^2$ from $z = 0$ to $z = 1 + i$ along the two different paths shown in Fig. 3.3.

(1) Along $y = x$ (shortest cut), $dy = dx$. Hence, we have

$$\int z^2 dz = \int (x^2 - y^2 + 2xyi)(dx + idy)$$

$$= \int_0^1 2x^2 i(1 + i)dx$$

$$= 2(i - 1)\int_0^1 x^2 dx$$

$$= \frac{2(i - 1)}{3}. \tag{3.4}$$

(2) Along $y = x^2$ and $dy = 2xdx$. Hence, we have

$$\int z^2 dz = \int (x^2 - y^2 + 2xyi)(dx + idy)$$

$$= \int_0^1 (x^2 - x^4 + 2x^3 i)(dx + i2xdx)$$

$$= \int_0^1 (x^2 - x^4 + 2x^3 i)(1 + 2ix)dx$$

$$= \frac{2(i - 1)}{3}. \tag{3.5}$$

This time, the integral values were the same. The function, $f(z)$, in Example 1 is not analytic but the function, $f(z)$, in Example 2 is analytic. This leads us to the following theorem.

Fig. 3.3 Integration of z^2 along two different paths

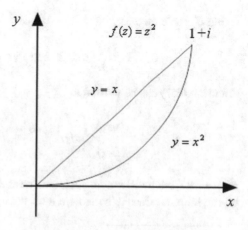

Fig. 3.4 Cauchy's theorem

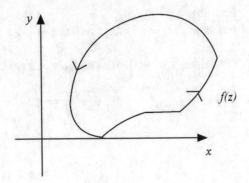

3.2   Cauchy's Theorem

Theorem 1 *If $f(z)$ is analytic inside and on a closed loop, C, as shown in Fig. 3.4, it[1]*
follows

$$\oint_C f(z)dz = 0. \tag{3.6}$$

Proof The contour integral can be expressed as the sum of two line integrals, one real and
the other imaginary as

$$\oint_C f(z)dz = \oint_C (u + iv)(dx + idy)$$

$$= \oint_C (udx - vdy) + i \oint_C (vdx + udy). \tag{3.7}$$

Using Green's theorem,

$$\oint_C (adx + bdy) = \int \int_D \left(\frac{\partial b}{\partial x} - \frac{\partial a}{\partial y}\right)dA. \tag{3.8}$$

Equation (3.7) can be written as

$$= -\int \int_D (v_x + u_y)dA + \int \int_D (u_x - v_y)dA$$

$$= 0, \tag{3.9}$$

where the Cauchy-Riemann equations were used. Note that if there is a singularity inside or
on the loop, C, Cauchy's theorem does not apply.   □

---

[1] $\oint_C f(z)dz$ denotes a contour integral of $f(z)$ along a loop C.

Fig. 3.5 Corollary of Cauchy's theorem

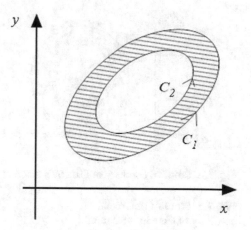

Corollary 1 *If $f(z)$ is analytic in the domain shown in Fig. 3.5 (but not necessarily analytic inside C_2), it follows*

$$\oint_{C_1} f(z)dz = \oint_{C_2} f(z)dz, \tag{3.10}$$

i.e., the contour of the integration can be deformed as long as it does not change the topology of singular points.

Proof If we choose a closed integral path shown in the left figure of Fig. 3.6 and assume that there are no singular points inside, the contour integral of $f(z)$ is 0 by Cauchy's theorem. Now, split the integral path into pieces as shown in the left figure of Fig. 3.6 by first traveling around the outermost segment, Γ_1, going inward along Γ_2, traveling along the inner segment, Γ_3 and finally returning to where it began along Γ_4. When we let the gap between Γ_2 and Γ_4 narrow and eventually to 0, the integrals along Γ_2 and Γ_4 are canceled each other because of the opposite signs. As the gap closes, the original contour integral ends up with two separate closed integrals along Γ_1 and Γ_3 as shown in the right figure of Fig. 3.6. Therefore, we have

$$\oint_{\Gamma_1} f(z)dz + \oint_{\Gamma_3} f(z)dz = 0, \tag{3.11}$$

or

$$\oint_{\Gamma_1} f(z)dz = \oint_{-\Gamma_3} f(z)dz, \tag{3.12}$$

or

$$\oint_{C_1} f(z)dz = \oint_{C_2} f(z)dz, \tag{3.13}$$

where $\Gamma_1 = C_1$ and $\Gamma_3 = -C_2$. □

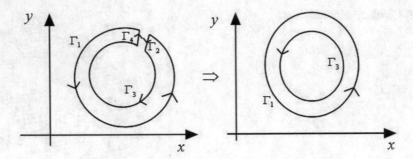

Fig. 3.6 Proof of corollary of Cauchy's theorem

Fig. 3.7 Proof of the second
corollary of Cauchy's theorem

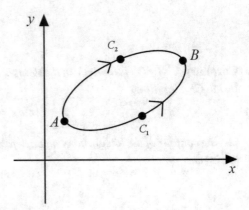

Corollary 2 *If $f(z)$ is analytic on and inside the domain as shown in Fig. 3.7, it follows*

$$\int_{A \to C_1 \to B} f(z)dz = \int_{A \to C_2 \to B} f(z)dz, \tag{3.14}$$

i.e., an integration of $f(z)$ from A to B is path independent.

Proof An integral of $f(z)$ along $A \to C_1 \to B \to C_2 \to A$ is zero according to Cauchy's theorem. Therefore,

$$\int_{A \to C_1 \to B \to C_2 \to A} f(z)dz = \int_{A \to C_1 \to B} f(z)dz + \int_{B \to C_2 \to A} f(z)dz$$

$$= \int_{A \to C_1 \to B} f(z)dz - \int_{A \to C_2 \to B} f(z)dz$$

$$= 0, \tag{3.15}$$

which is equivalent to

$$\int_{A \to C_1 \to B} f(z)dz = \int_{A \to C_2 \to B} f(z)dz. \tag{3.16}$$

$\square$

Back to the examples of complex integrals for z^2 and $|z|^2$, it is seen from this result that an integration of z^2 is independent of the integral path as z^2 is analytic everywhere while an integration of $|z|^2$ generally depends on the integral path as Corollary 1 does not apply to $|z|^2$.

3.2.1 Morera's Theorem

Theorem 2 *If $f(z)$ is continuous in a simply connected domain, D, and*

$$\oint_C f(z) = 0, \tag{3.17}$$

along any closed contour, C, in D, then $f(z)$ is analytic in D.

Proof Let $F(z)$ be defined as

$$F(z) \equiv \int_{z_0}^{z} f(z)dz, \tag{3.18}$$

where z_0 is an arbitrary point in D. From Corollary 2, $F(z)$ is independent of the integral path. Hence, we have

$$F(z + \Delta z) - F(z) = \int_{z}^{z+\Delta z} f(z)dz \sim f(z)\Delta z. \tag{3.19}$$

By dividing both sides of Eq. (3.19) by Δz and letting $\Delta z \to 0$, we have

$$F'(z) = f(z). \tag{3.20}$$

Therefore, $F(z)$ is analytic and can be differentiated, which is $f(z)$. Hence, $f(z)$ is analytic at z_0.
$\square$

Morera's theorem is the inverse of Cauchy's theorem. It is useful when we want to prove that $f(z)$ is analytic [3].

3.3   Cauchy's Integral Formula

3.3.1   Contour Integral of z^n

Consider a contour integral of z^n where the contour is a closed loop containing $z = 0$ as shown in Fig. 3.8. The following result is important and will be used in the subsequent development.

$$\oint_C z^n dz = \begin{cases} 2\pi i, \ n = -1 \\ 0, \quad \text{otherwise.} \end{cases} \tag{3.21}$$

Proof For $n = 0, 1, 2, \ldots, z^n$ is analytic everywhere, hence,

$$\oint_C z^n dz = 0, \tag{3.22}$$

thanks to Cauchy's theorem.

For $n = -1$,

$$\oint_C \frac{dz}{z} = \left[\log z\right]_{z=e^0}^{z=e^{2\pi i}}$$

$$= \ln e^{2\pi i} - \ln e^0$$

$$= 2\pi i. \tag{3.23}$$

For $n = -m, \quad m \geq 2$,

Fig. 3.8 Contour Integral of z^n

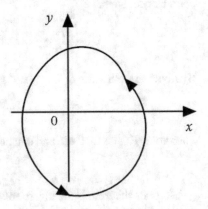

$$\oint_C \frac{dz}{z^m} = \frac{1}{1-m}\left[\frac{1}{z^{m-1}}\right]_{z=e^0}^{z=e^{2\pi i}}$$

$$= \frac{1}{1-m}\left(\frac{1}{e^{2\pi(m-1)i}} - \frac{1}{e^0}\right)$$

$$= 0. \tag{3.24}$$

This integral is 0 not because of Cauchy's theorem but because of the direct integration. □

Variation

By shifting the origin to a as in Fig. 3.9, Eq. (3.21) can be modified as

$$\int_C (z-a)^n dz = \begin{cases} 2\pi i, & n = -1 \\ 0, & \text{otherwise,} \end{cases} \tag{3.25}$$

where C is a loop which contains $z = a$.

3.3.2   Cauchy's Integral Formula

If $f(z)$ is analytic on and inside a closed loop, C, and a is a point inside the loop as in Fig. 3.9, it follows

$$f(a) = \frac{1}{2\pi i}\oint_C \frac{f(z)}{z-a}dz, \tag{3.26}$$

i.e., the values of $f(z)$ inside the loop are completely determined from the values of $f(z)$ on the loop alone. Equation (3.26) is called Cauchy's integral formula.

Proof First, according to Eq. (3.10), the integral path can be shrank to a circle centered at a with a radius ϵ. Thus, the contour integral of $f(z)$ along C can be written as

Fig. 3.9 Contour integral
around $z = a$

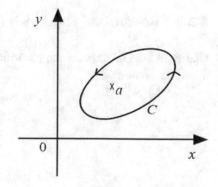

$$\oint_C \frac{f(z)}{z-a}dz = \oint_{|z-a|=\epsilon} \frac{f(z)}{z-a}dz$$

$$= \oint_{|z-a|=\epsilon} \frac{f(z) - f(a) + f(a)}{z-a}dz$$

$$= \oint_{|z-a|=\epsilon} \frac{f(z) - f(a)}{z-a}dz + \oint_{|z-a|=\epsilon} \frac{f(a)}{z-a}dz. \qquad (3.27)$$

In the first integral, we can set $z - a = \epsilon e^{i\theta}, dz = \epsilon i e^{i\theta} d\theta$, so that the first integral becomes

$$\oint_{|z-a|=\epsilon} \frac{f(z) - f(a)}{z-a}dz = \int_0^{2\pi} \frac{f(a + \epsilon e^{i\theta}) - f(a)}{\epsilon e^{i\theta}} i\epsilon e^{i\theta} d\theta$$

$$= i \int_0^{2\pi} \left(f(a + \epsilon e^{i\theta}) - f(a) \right) d\theta$$

$$\to 0 \quad \text{as } \epsilon \to 0, \qquad (3.28)$$

because $f(z)$ is analytic at $z = a$.

The second integral is evaluated as

$$\oint_{|z-a|=\epsilon} \frac{f(a)}{z-a}dz = f(a) \oint_{|z-a|=\epsilon} \frac{1}{z-a}dz = 2\pi i f(a). \qquad (3.29)$$

Therefore, we have

$$\oint_C \frac{f(z)}{z-a}dz = 2\pi i f(a). \qquad (3.30)$$

$\square$

Cauchy's integral formula implies that once an analytic function, $f(z)$, is defined on a closed loop alone, the values of $f(z)$ at all the points inside the loop are automatically determined. This again underscores the fact that analytic functions are restricted and exclusive.

3.3.3 Generalized Cauchy's Integral Formula

Cauchy's integral formula can be further extended to express $f'(a)$ by the values of $f(z)$ along the contour as

$$f'(a) = \frac{1}{2\pi i} \oint \frac{f(z)}{(z-a)^2}dz. \qquad (3.31)$$

Proof From Cauchy's integral formula, we have

$$f(a + \Delta a) = \frac{1}{2\pi i} \oint \frac{f(z)}{z - (a + \Delta a)} dz, \tag{3.32}$$

$$f(a) = \frac{1}{2\pi i} \oint \frac{f(z)}{z - a} dz. \tag{3.33}$$

Subtracting Eq. (3.33) from Eq. (3.32) and dividing the result by Δa yield

$$\frac{f(a + \Delta a) - f(a)}{\Delta a} = \frac{1}{2\pi i} \oint f(z) \left(\frac{1}{z - (a + \Delta a)} - \frac{1}{z - a} \right) dz$$

$$= \frac{1}{2\pi} \oint f(z) \left(\frac{1}{(z - a)(z - a - \Delta a)} \right) dz. \tag{3.34}$$

By letting $\Delta a \to 0$, we have

$$f'(a) = \frac{1}{2\pi i} \oint \frac{f(z)}{(z - a)^2} dz. \tag{3.35}$$

$\square$

This result can be further generalized to yield the following formula known as *generalized Cauchy's integral formula*.

$$f^{(n)}(a) = \frac{n!}{2\pi i} \oint_C \frac{f(z)}{(z - a)^{n+1}} dz. \tag{3.36}$$

We observe that once the values of $f(z)$ on its contour are prescribed over a loop, $f(z)$ is uniquely determined in the loop. This statement is further extended that once the values of $f(z)$ on its contour are prescribed, not only $f(z)$ but also $f'(z)$, $f''(z)$ and subsequent higher order derivatives are uniquely determined inside the loop. We can thus state that an analytic function, $f(z)$, can be differentiated an infinite number of times. This again underscores the fact that analyticity of $f(z)$ is such a strong restriction.[2]

---

[2] A quick (and unofficial) way to derive Cauchy's integral formula (Eq. (3.36)) is to start writing the Taylor series of $f(z)$ about $z = a$ as

$$f(z) = f(a) + f'(a)(z - a) + \frac{f''(a)}{2!}(z - a)^2 + \ldots + \frac{f^{(n)}(a)}{n!}(z - a)^n + \ldots. \tag{3.37}$$

By dividing both sides of Eq. (3.37) by $(z - a)^{n+1}$, we have

$$\frac{f(z)}{(z - a)^{n+1}} = \frac{f(a)}{(z - a)^{n+1}} + \frac{f'(a)}{(z - a)^n} + \frac{f''(a)}{2!} \frac{1}{(z - a)^{n-1}} + \ldots + \frac{f^{(n)}(a)}{n!} \frac{1}{z - a} + \ldots. \tag{3.38}$$

By integrating both sides of Eq. (3.38) along a loop that contains $z = a$, we have

Fig. 3.10 Contour integral
with $z = 3$ as a singular point
inside

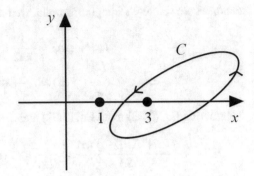

Example 1

Evaluate

$$I = \oint_C \frac{e^z}{(z-1)(z-3)} dz, \tag{3.41}$$

where C is a loop shown in Fig. 3.10. Even though both $z = 1$ and $z = 3$ make the denominator vanish, only $z = 3$ is inside C.

If we define

$$f(z) \equiv \frac{e^z}{z-1}, \tag{3.42}$$

$f(z)$ is analytic on and inside the loop, C. Thereby,

$$I = \oint_C \frac{f(z)}{z-3} dz$$
$$= 2\pi i f(3)$$
$$= 2\pi i \frac{e^3}{3-1}$$
$$= \pi e^3 i. \tag{3.43}$$

$$\oint_C \frac{f(z)}{(z-a)^{n+1}} dz = f(a) \oint_C \frac{dz}{(z-a)^{n+1}} + f'(a) \oint_C \frac{dz}{(z-a)^n}$$
$$+ \frac{f''(a)}{2!} \oint_C \frac{dz}{(z-a)^{n-1}} + \ldots + \frac{f^{(n)}(a)}{n!} \oint_C \frac{dz}{z-a} + \ldots$$
$$= f(a) \times 0 + f'(a) \times 0 + \frac{f''(a)}{2!} \times 0 + \ldots + \frac{f^{(n)}(a)}{n!} \times 2\pi i + \ldots$$
$$= 2\pi i \frac{f^{(n)}(a)}{n!}, \tag{3.39}$$

thus, we have

$$f^{(n)}(a) = \frac{n!}{2\pi i} \oint_C \frac{f(z)}{(z-a)^{n+1}} dz. \tag{3.40}$$

This is not a proof as we have not defined the Taylor series for complex functions yet until Sect. 4.1. Nevertheless, it provides a quick way to write down Cauchy's integral formula.

Example 2

Same as Example 1 but for the loop, C, shown in Fig. 3.11.

In this case, the loop C contains both $z = 1$ and $z = 3$. However, it is possible to deform the loop as long as the loop can contain the two points. Following Fig. 3.12, the original path can be deformed to another contour shown in the second figure of Fig. 3.12 without changing the value of the integral.

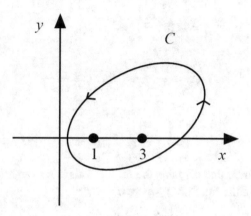

Fig. 3.11 Contour integral with $z = 1$ and $z = 3$ as singular points inside

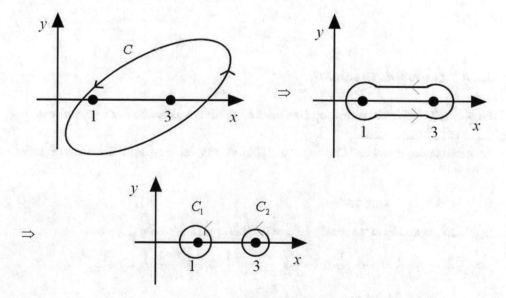

Fig. 3.12 Contour integral to separate two contour integrals

With the deformation shown in the third figure of Fig. 3.12, the integration can be split into two independent integrals around $z = -4$ and $z = 2$ as

$$I = \oint_{C_1} \frac{f_1(z)}{z - 1} dz + \oint_{C_2} \frac{f_2(z)}{z - 3} dz$$
$$= 2\pi i f_1(1) + 2\pi i f_2(3)$$
$$= 2\pi i \frac{e}{1 - 3} + 2\pi i \frac{e^3}{3 - 1}$$
$$= i e \left(e^2 - 1 \right) \pi, \tag{3.44}$$

where

$$f_1(z) \equiv \frac{e^z}{z - 3}, \quad f_2(z) \equiv \frac{e^z}{z - 1}. \tag{3.45}$$

Example 3

Evaluate

$$I \equiv \oint_C \frac{e^z}{z^4} dz, \tag{3.46}$$

where C is the unit circle that contains the origin. This is a straightforward application of Cauchy's integral formula, Eq. (3.36), where $n = 3$.[3]

$$\oint \frac{e^z}{z^4} dz = \frac{2\pi i}{3!} f'''(0)$$
$$= \frac{\pi i}{3}. \tag{3.49}$$

3.3.4   Liouville's Theorem

Theorem 3 *If $f(z)$ is analytic and bounded on the entire plane, it must be a constant.*

---

[3] Instead of using generalized Cauchy's formula (Eq. (3.36)), we can expand e^z directly by the Taylor series as

$$e^z = 1 + z + \frac{z^2}{2!} + \frac{z^3}{3!} + \frac{z^4}{4!} + \dots \tag{3.47}$$

By dividing both sides of the above by z^4 and integrating the result along C, we have

$$\oint_C \frac{e^z}{z^4} dz = \oint_C \frac{1}{z^4} dz + \oint_C \frac{1}{z^3} dz + \frac{1}{2!} \oint_C \frac{1}{z^2} dz + \frac{1}{3!} \oint_C \frac{1}{z} dz + \frac{1}{4!} \oint_C 1 dz + \dots$$
$$= 0 + 0 + 0 + \frac{1}{3!} \times 2\pi i + 0 + \dots$$
$$= \frac{\pi i}{3}. \tag{3.48}$$

This is, of course, based on acceptance of the Taylor series for complex functions.

Fig. 3.13 Contour integral for Liouville's theorem

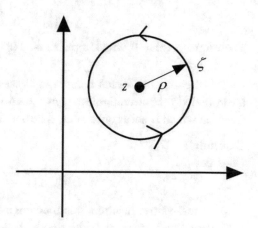

Proof By changing a to z and z to ζ in Eq. (3.35) as shown in Fig. 3.13, we have

$$f'(z) = \frac{1}{2\pi i} \oint_C \frac{f(\zeta)}{(\zeta - z)^2} d\zeta. \tag{3.50}$$

Choose C as a circle centered at z with a radius ρ. Let M be the maximum value of $f(\zeta)$ on C. Then, we have

$$
\begin{aligned}
\left| f'(z) \right| &= \left| \frac{1}{2\pi i} \oint_C \frac{f(\zeta)}{(\zeta - z)^2} d\zeta \right| \\
&= \frac{1}{2\pi} \left| \oint_C \frac{f(\zeta)}{(\zeta - z)^2} d\zeta \right| \\
&\leq \frac{1}{2\pi} \oint_C \left| \frac{f(\zeta)}{(\zeta - z)^2} \right| d\zeta \\
&\leq \frac{1}{2\pi} \frac{M}{\rho^2} \oint_C d\zeta \\
&\leq \frac{1}{2\pi} \frac{M}{\rho^2} \oint_C |d\zeta| \\
&= \frac{1}{2\pi} \frac{M}{\rho^2} 2\pi\rho \\
&= \frac{M}{\rho}, \tag{3.51}
\end{aligned}
$$

where we used

$$\left| \oint g(\zeta) d\zeta \right| \leq \oint |g(\zeta)| d\zeta. \tag{3.52}$$

As M in Eq. (3.51) remains finite while ρ in Eq. (3.51) can be chosen arbitrarily large, it follows

$$f'(z) \to 0 \quad \text{as} \quad \rho \to \infty. \tag{3.53}$$

Therefore, $f'(z) = 0$ which implies that $f(z) = const$.   □

This theorem first seems to defy intuition as in real variables, many real-valued functions are bounded yet differentiable everywhere. According to the Liouville theorem, any analytic function which is not a constant must be unbounded at some point(s) on the complex plane.

Example 1

Consider

$$f(x) = \frac{1}{1 + x^2}. \tag{3.54}$$

As a real-valued function, this function is bounded as seen in Fig. 3.14. However, as a complex-valued function, $f(z) \to \infty$ as $z \to \pm i$.

Example 2

Consider

$$f(x) = \cos x. \tag{3.55}$$

This function, as a real-valued function, is bounded between -1 and 1. However, as a complex-valued function, it becomes unbounded as the imaginary part of z goes to $\pm \infty$ as shown in Fig. 3.15.

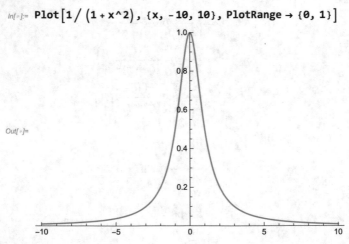

In[ ]:= **Plot$\left[1 / \left(1 + x^\wedge 2\right), \{x, -10, 10\}, \text{PlotRange} \to \{0, 1\}\right]$**

Out[ ]=

Fig. 3.14 Graph of $\frac{1}{1+x^2}$

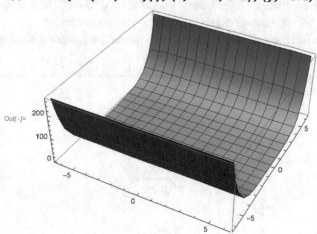

$In[\circ]:=$ `Plot3D[Abs[Cos[x + I y]], {x, -2 Pi, 2 Pi}, {y, -2 Pi, 2 Pi}]`

$Out[\circ]=$

Fig. 3.15 3-D plot of $|\cos(x + iy)|$

3.3.5 Fundamental Theorem of Algebra

Theorem 4 *An n-th order algebraic equation,* $P_n(z) = 0$, *has at least one root where*

$$P_n(z) \equiv a_n z^n + a_{n-1} z^{n-1} + \ldots a_1 z + a_0. \tag{3.56}$$

Proof Let

$$f(z) \equiv \frac{1}{P(z)}. \tag{3.57}$$

If we assume that $P_n(z)$ is never equal to 0 (i.e., $P_n(z)$ has no zeros), then $f(z)$ is analytic everywhere (the denominator never goes to 0) and it is bounded ($P(z) \to \infty$ as $z \to \infty$ along the real axis). Thus, according to the Liouville theorem, $P(z)$ must be a constant, which is not the case; hence, the assumption ($P_n(z)$ never goes to 0) is wrong.[4]   □

If $P_n(z)$ has a zero, say, a_1, it can be factorized as

$$P_n(z) = (z - a_1)(b_{n-1} z^{n-1} + b_{n-2} z^{n-2} + \ldots b_1 z + b_0). \tag{3.58}$$

Applying the fundamental theorem of algebra recursively, it is seen that an n-th order algebraic equation has n roots (including multiple roots).

A fundamental question would have been that if the complete solution for quadratic equations requires $i = \sqrt{-1}$, isn't it necessary that we need to introduce another element,

---

[4] Unfortunately, the fundamental theorem of algebra does not tell how to actually solve the algebraic equation.

j, for the third-order equations, k for quartic equations and so on? The answer is that once i is introduced, it will take care of algebraic equations of any order.

The fundamental theorem of algebra can be also proven using the argument principle [1].

3.4   Problems

1. Evaluate the following complex integral:

$$\oint_{|z|=2} \frac{\cos z}{z^3(z^2 + 1)} dz. \tag{3.59}$$

2. Evaluate

$$\oint \frac{1}{z - 1} dz, \tag{3.60}$$

   where the integral path is a rectangle whose corners are $-1, +2, 2 + i$ and $-1 + i$.

3. Evaluate

$$\int_1^{1+i} \sin z \, dz, \tag{3.61}$$

   where the integral path is the straight line from 1 to $1 + i$.

4. Evaluate

$$\int_C \left(-2xy + i(x^2 - y^2)\right) dx + \left(y^2 - x^2 - 2ixy\right) dy, \tag{3.62}$$

   where C is a curve $y = x^2, 0 \le x \le 1$.

Series of Complex Variable Functions

4

In this chapter, we will show that any analytic functions can be expressed as infinite series. The convergence of such series depends on the location of singularities and the center of expansion. In fact, infinite series is the standard form of analytic functions, and if we can sum up an infinite series to a closed form, we are extremely lucky!

In real variables, only the Taylor series was relevant but in complex variables, depending on the distribution of singular points, the Laurent series expansion must be employed in addition to the Taylor series. The Taylor series is not fully understood until the Taylor series for complex functions is looked upon.

An analytic function expressed in an infinite series in a limited region can be extended to the region outside by analytic continuation, effectively expanding its definition to cover the entire plane except for singular points.

4.1 Taylor Series

To show how an analytic function can be expanded by an infinite series in different ways, consider expanding $f(z) = 1/(1 + z)$ about $z = 0$. If we accept that the geometric series for real functions can be used for complex functions, $f(z)$ is expressed as

$$\frac{1}{1+z} = 1 - z + z^2 - z^3 + \cdots . \tag{4.1}$$

S. Nomura, *Complex Variables for Engineers with Mathematica*,
Synthesis Lectures on Mechanical Engineering,
https://doi.org/10.1007/978-3-031-13067-0_4

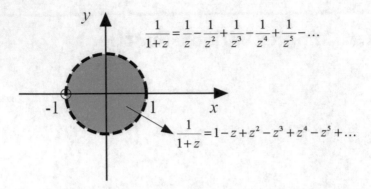

Fig. 4.1 Two different ways to expand $\frac{1}{1+z}$ about $z = 0$

From the ratio test[1] of convergence, Eq. (4.1) is convergent if $|z| < 1$. However, the same function can be also expanded as

$$\frac{1}{1+z} = \frac{1}{z(1+\frac{1}{z})}$$

$$= \frac{1}{z}\left(1 - \frac{1}{z} + \left(\frac{1}{z}\right)^2 - \left(\frac{1}{z}\right)^3 + \cdots\right). \tag{4.3}$$

The range of convergence for Eq. (4.3) is $|z| > 1$ by the ratio test. It is seen that both Eqs. (4.1) and (4.3) are partial representations of $f(z) = 1/(1 + z)$ as shown in Fig. 4.1. The series, Eq. (4.1), is the Taylor series for $f(z)$ about $z = 0$ and the series, Eq. (4.3), is the Laurent series of $f(z)$ about $z = 0$ in the two mutually exclusive regions.

4.1.1 Taylor Series of $f(z)$ About $z = a$

It is shown that the Taylor series for complex functions is derived directly from Cauchy's integral formula without reference to the Taylor series for real functions. Deriving the Taylor series from Cauchy's integral formula also reveals the limitation of the Taylor series.

---

[1] An infinite series,

$$\sum a_n = a_0 + a_1 + a_2 + a_3 + \cdots,$$

converges if

$$\lim_{n\to\infty}\left|\frac{a_{n+1}}{a_n}\right| = r,$$

and

$$r < 1. \tag{4.2}$$

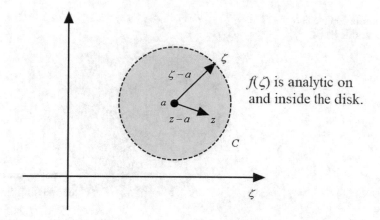

Fig. 4.2 A circle centered at $\zeta = a$ and z inside

Using ζ as the active variable, assume that $f(\zeta)$ is analytic on and inside a circular disk, C, as shown in Fig. 4.2. The value of $f(z)$ at $\zeta = z$ inside the disk can be expressed by Cauchy's integral formula as

$$f(z) = \frac{1}{2\pi i} \oint_C \frac{f(\zeta)}{\zeta - z} d\zeta. \tag{4.4}$$

Noting that $|\zeta - a| > |z - a|$, Eq. (4.4) can be rewritten as

$$
\begin{aligned}
f(z) &= \frac{1}{2\pi i} \oint_C \frac{f(\zeta)}{\zeta - z} d\zeta \\
&= \frac{1}{2\pi i} \oint_C \frac{f(\zeta)}{(\zeta - a)(1 - \frac{z-a}{\zeta - a})} d\zeta \\
&= \frac{1}{2\pi i} \oint_C \frac{f(\zeta)}{\zeta - a} \left(1 + \left(\frac{z-a}{\zeta - a} \right) + \left(\frac{z-a}{\zeta - a} \right)^2 + \left(\frac{z-a}{\zeta - a} \right)^3 + \cdots \right) d\zeta \\
&= \left(\frac{1}{2\pi i} \oint_C \frac{f(\zeta)}{\zeta - a} d\zeta \right) + \left(\frac{1}{2\pi i} \oint_C \frac{f(\zeta)}{(\zeta - a)^2} d\zeta \right) (z - a) \\
&\quad + \left(\frac{1}{2\pi i} \oint_C \frac{f(\zeta)}{(\zeta - a)^3} d\zeta \right) (z - a)^2 + \left(\frac{1}{2\pi i} \oint_C \frac{f(\zeta)}{(\zeta - a)^4} d\zeta \right) (z - a)^3 + \cdots \\
&= f(a) + f'(a)(z - a) + \frac{f''(a)}{2!} (z - a)^2 + \frac{f'''(a)}{3!} (z - a)^3 + \cdots, \tag{4.5}
\end{aligned}
$$

where we used generalized Cauchy's integral formula (Eq. (3.36)) to express each contour integral on the fourth and fifth lines of Eq. (4.5) by the corresponding derivatives of $f(a)$.

As expected, this result is identical to the Taylor series for real functions. However, noting that this was derived based on Cauchy's integral formula, the validity of the Taylor series in

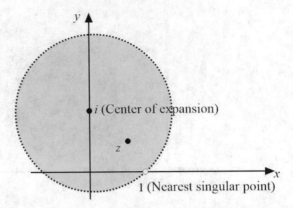

Fig. 4.3 A circle centered at $z = i$ with a radius of $\sqrt{2}$

i (Center of expansion)

z

1 (Nearest singular point)

complex variables is now made clear, i.e., the expansion above is valid until the disk expands to a region where $f(z)$ is no longer analytic, or until the disk hits the first nearest singular point(s) of $f(z)$.

Example 1

Expand $f(z) = 1/(z - 1)$ by the Taylor series about $z = i$ in the shaded region shown in Fig. 4.3.

Note that the expansion must be in the format of

$$f(z) = \sum c_n(z - i)^n, \tag{4.6}$$

as $z = i$ is the center of expansion. The valid range of the Taylor series for $1/(z - 1)$ is within the circle centered at $z = i$ with a radius of $\sqrt{2}(= |i - 1|)$ as $z = 1$ is the nearest singular point from $z = i$. We could always use Eq. (4.5), the formula for the Taylor series, but a direct expansion method works better in many cases. By extracting the term $(z - i)$ from $(z - 1)$, we have

$$\frac{1}{z - 1} = \frac{1}{(z - i) - (1 - i)}$$

$$= \frac{1}{-(1 - i)(1 - \frac{z-i}{1-i})}$$

$$= \frac{1}{-(1 - i)} \left(1 + \left(\frac{z - i}{1 - i} \right) + \left(\frac{z - i}{1 - i} \right)^2 + \left(\frac{z - i}{1 - i} \right)^3 + \cdots \right). \tag{4.7}$$

The range of convergence is

$$|z - i| < \sqrt{2}. \tag{4.8}$$

In the derivation above, we could use $1 - i$ as the pivot instead of $z - i$. However, by doing so, we have a term

Fig. 4.4 Two different zones
for a circle centered at $z = 0$

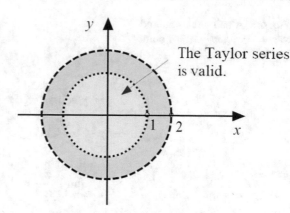

The Taylor series
is valid.

$$\frac{1}{1 - \frac{1-i}{z-i}} \qquad (4.9)$$

which allows the geometric series expansion with negative powers of $\frac{1}{z-i}$, which is not the
Taylor series.

Example 2

Expand

$$f(z) \equiv \frac{1}{z^2 - 3z + 2}, \qquad (4.10)$$

by the Taylor series about $z = 0$.

As shown in Fig. 4.4, $f(z)$ can be expanded by the Taylor series inside (but not on) the
disk, $|z| < 1$, as there is no singular point inside the disk. Thus, we have

$$\frac{1}{z^2 - 3z + 2} = \frac{1}{z-2}\frac{1}{z-1}$$

$$= \frac{1}{(-2)(1 - z/2)}\frac{1}{(-1)(1 - z)}$$

$$= \frac{1}{(-2)}\left(1 + \frac{z}{2} + \left(\frac{z}{2}\right)^2 + \left(\frac{z}{2}\right)^3 + \cdots\right) \times$$

$$\frac{1}{(-1)}\left(1 + z + z^2 + z^3 + z^4 + \cdots\right). \qquad (4.11)$$

Example 3

Expand

$$f(z) \equiv \frac{z}{\sin z}, \qquad (4.12)$$

about $z = 0$.

Note that $z = 0$ is not a singular point as

Fig. 4.5 A unit circle centered at $z = 0$ with no singular point inside

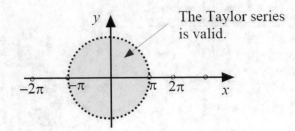

The Taylor series is valid.

$$\lim_{z \to 0} \frac{z}{\sin z} = 1. \tag{4.13}$$

The zeros of $\sin z$ are $z = n\pi$ where n is an integer (but $n \neq 0$) that are distributed on the real axis as shown in Fig. 4.5. As the nearest singular points from $z = 0$ are $z = \pm\pi$, $f(z)$ can be expanded by the Taylor series in $|z| < \pi$. Knowing that $f(z)$ is Taylor series expandable, you can use the formula directly, i.e.,

$$f(z) = f(0) + f'(0)z + \frac{f''(0)}{2!}z^2 + \frac{f'''(0)}{3!}z^3 + \cdots, \tag{4.14}$$

where

$$f(0) = \lim_{z \to 0} \frac{z}{\sin z} = 1, \tag{4.15}$$

$$f'(0) = \lim_{z \to 0} f'(z)$$
$$= \lim_{z \to 0} \frac{\sin z - z \cos z}{\sin^2 z}$$
$$= \cdots$$
$$= 0, \tag{4.16}$$

$$f''(0) = \cdots$$
$$= \frac{1}{3}. \tag{4.17}$$
$$\cdots$$

Therefore,

$$\frac{z}{\sin z} = 1 + \frac{z^2}{6} + \frac{7z^4}{360} + \cdots. \tag{4.18}$$

You can also try long division as

$$z - z^3/6 + z^5/120 - \cdots \frac{1 + \frac{z^2}{6} + \quad \cdots}{\sqrt{z}}$$

$$\frac{z - z^3/6 + z^5/120 - \cdots}{z^3/6 - z^5/120 + \cdots} \tag{4.19}$$

$$\frac{z^3/6 - z^5/36 + \cdots}{7z^5/360 - \cdots}$$

Mathematica **Code**

Mathematica has a built-in function, `Series[f[z]]`, which returns the Taylor series of a complex function, `f[z]`. The function, `Series[f[z], z, a, n]`, returns the Taylor series of $f(z)$ about $z = a$ up to the n-th order. For example, the Taylor series of $\frac{e^z}{1-z}$ about $z = 0$ up to the seventh order can be computed as

```
In[ ]:= Series[Exp[z] / (1 - z), {z, 0, 7}]
```

$$Out[ ]= 1 + 2z + \frac{5z^2}{2} + \frac{8z^3}{3} + \frac{65z^4}{24} + \frac{163z^5}{60} + \frac{1957z^6}{720} + \frac{685z^7}{252} + O[z]^8$$

The last term in the output above is the order of the remainder. The output can be converted to a normal expression using `Normal[]` as

```
In[ ]:= Normal[%]
```

$$Out[ ]= 1 + 2z + \frac{5z^2}{2} + \frac{8z^3}{3} + \frac{65z^4}{24} + \frac{163z^5}{60} + \frac{1957z^6}{720} + \frac{685z^7}{252}$$

The Taylor series for Example 3 above can be obtained as

```
In[ ]:= Series[z / Sin[z], {z, 0, 10}]
```

$$Out[ ]= 1 + \frac{z^2}{6} + \frac{7z^4}{360} + \frac{31z^6}{15\,120} + \frac{127z^8}{604\,800} + \frac{73z^{10}}{3\,421\,440} + O[z]^{11}$$

Remark on Taylor Series

Consider the following geometric series.

$$\frac{1}{1 - z} = 1 + z + z^2 + z^3 + z^4 + \cdots . \tag{4.20}$$

By the ratio test, $|z^{n+1}/z^n| = |z| < 1$ must be satisfied for the convergence of the series which makes sense because both sides of Eq. (4.20) diverge if we substitute z for 1.

Now consider

$$\frac{1}{1 + z^2} = 1 - z^2 + z^4 - z^6 + z^8 - \cdots . \tag{4.21}$$

By the ratio test, $|z^{2n+2}/z^{2n}| = |z^2| < 1$. Therefore, the range of the convergence is $|z| < 1$. However, $1/(1 + z^2)$ is evaluated to be $\frac{1}{2}$ as $z \to 1$ unlike $\frac{1}{1-z}$. This seemingly odd result cannot be explained if we restrict ourselves to real-valued functions.

Fig. 4.6 Singular points on the
imaginary axis for $\frac{1}{1+z^2}$

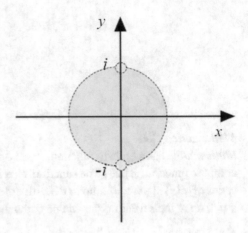

Note that the function, $f(z) = 1/(1 + z^2)$, has singular points at $z = \pm i$ but there is no singular point on the real axis as in Fig. 4.6. The function, $f(z)$, is actually not singular at $z = 1$ even though $z = 1$ lies on $|z| = 1$. In fact, $|z| < 1$ is a sufficient condition for $f(z)$ to be analytic but not a necessary condition as there may be a point in $|z| \geq 1$ at which $f(z)$ is still analytic.

4.1.2 Analytic Continuation

To elucidate the concept of analytic continuation, consider

$$f(z) = \frac{1}{2}\left(1 - \frac{z}{2} + \left(\frac{z}{2}\right)^2 - \left(\frac{z}{2}\right)^3 + \cdots\right). \tag{4.22}$$

From the ratio test, the range of convergence is $|z| < 2$, which implies that $f(3)$ (outside the disk) cannot be computed. On the other hand, as $z = 1$ is within the disk, it is possible to compute $f(1), f'(1), f''(1) \cdots$ as

$$f(1) = \frac{1}{2}\left(1 - \frac{1}{2} + \left(\frac{1}{2}\right)^2 - \cdots\right)$$

$$= \frac{1}{2}\frac{1}{1 + \frac{1}{2}}$$

$$= \frac{1}{3}. \tag{4.23}$$

For the derivatives,

$$f'(z) = \frac{1}{2}\left(0 - \frac{1}{2} + 2\left(\frac{z}{2}\right)\frac{1}{2} - 3\left(\frac{z}{2}\right)^2\frac{1}{2} + \cdots\right)$$
$$= \left(\frac{-1}{4}\right)\left(1 - 2\left(\frac{z}{2}\right) + 3\left(\frac{z}{2}\right)^2 - 4\left(\frac{z}{2}\right)^3 + \cdots\right). \tag{4.24}$$

Hence,[2]

$$f'(1) = \frac{-1}{4}\left(1 - 2\left(\frac{1}{2}\right) + 3\left(\frac{1}{2}\right)^2 - 4\left(\frac{1}{2}\right)^3 + \cdots\right)$$
$$= \left(-\frac{1}{4}\right)\frac{1}{\left(1 + \frac{1}{2}\right)^2}$$
$$= -\frac{1}{9}. \tag{4.28}$$

Similarly,

$$f''(z) = \frac{1}{2}\left(\frac{1}{2} - \frac{3}{4}z + \frac{3}{4}z^2 - \frac{5}{8}z^3 + \cdots\right), \tag{4.29}$$

and

$$f''(1) = \frac{1}{2}\left(\frac{1}{2} - \frac{3}{4} + \frac{3}{4} - \frac{5}{8} + \cdots\right) = \frac{2}{27}. \tag{4.30}$$

With the values of $f(1)$, $f'(1)$, $f''(1)$, ..., we can define a new function, $g(z)$, as

$$g(z) \equiv f(1) + f'(1)(z - 1) + \frac{f''(1)}{2!}(z - 1)^2 + \cdots$$
$$= \frac{1}{3} - \frac{1}{9}(z - 1) + \frac{1}{27}(z - 1)^2 - \cdots. \tag{4.31}$$

It is noted that $g(z)$ is identical to $f(z)$ in $|z| < 1$ as it is the Taylor series of $f(z)$ about $z = 1$. Nevertheless, from the ratio test, the range of convergence for $g(z)$ is found to be

$$\left|\frac{\frac{1}{27}(z - 1)^2}{\frac{1}{9}(z - 1)}\right| = \frac{|z - 1|}{3} < 1, \tag{4.32}$$

2

$$1 - \alpha + \alpha^2 - \alpha^3 + \cdots = \frac{1}{1 + \alpha}. \tag{4.25}$$

Differentiating the above with respect to α yields

$$-1 + 2\alpha - 3\alpha^2 + 4\alpha^3 - \cdots = -\frac{1}{(1 + \alpha)^2}, \tag{4.26}$$

or

$$1 - 2\alpha + 3\alpha^2 - 4\alpha^3 + \cdots = \frac{1}{(1 + \alpha)^2}. \tag{4.27}$$

Fig. 4.7 Analytic continuation

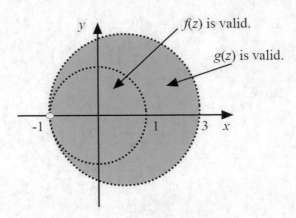

i.e.,

$$|z - 1| < 3. \tag{4.33}$$

The range of convergence for $g(z)$ expressed by Eq. (4.33) contains the range of convergence for $f(z)$, $|z| < 2$, yet $g(z)$ is identical to $f(z)$ in $|z| < 2$ so we effectively extended the definition of $f(z)$ from $|z| < 2$ to $|z - 1| < 3$ as shown in Fig. 4.7. This process can be repeated until the entire plane is covered except for $z = -2$ and is called *analytic continuation*.

Theorem 5 *If $F(z)$ is analytic in a domain, D, and $F(z) = 0$ on some open disk in D, then $F(z) = 0$ throughout D.*

Proof If $F(z) = 0$ in that disk, all the Taylor series coefficients are 0 in the disk. Using analytic continuation, all the Taylor series coefficients are 0 in D. □

Identity Theorem

Theorem 6 *If $f(z)$ and $g(z)$ are analytic in a domain, D, and $f(z) \equiv g(z)$ in some disk in D, then $f(z) = g(z)$ throughout D.*

Proof Set $F(z) \equiv f(z) - g(z)$ in Theorem 5. □

Example
Prove that

$$\sin^2 z + \cos^2 z = 1, \tag{4.34}$$

for all $z = x + yi$.

1. Long answer:

$$\cos z = \frac{1}{2}\left(e^{iz} + e^{-iz}\right)$$
$$= \frac{1}{2}\left(e^{i(x+iy)} + e^{-i(x+iy)}\right)$$
$$= \cdots$$
$$= \cos x \cosh y - i \sin x \sinh y. \qquad (4.35)$$

Similarly,

$$\sin z = \sin x \cosh y + i \cos x \sinh y. \qquad (4.36)$$

Therefore,

$$\sin^2 z + \cos^2 z = (\cos x \cosh y - i \sin x \sinh y)^2 + (\sin x \cosh y + i \cos x \sinh y)^2$$
$$= \cdots$$
$$= 1. \qquad (4.37)$$

2. Short answer:
   Define $f(z) = \sin^2 z + \cos^2 z$ and $g(z) \equiv 1$. Then, $f = g$ for real z. Hence, $f = g$ for all z per Theorem 6.

4.1.3 Can We Prove $1 + 2 + 3 + 4 + \cdots = -\frac{1}{12}$?

The sum of all the natural numbers,

$$\sum_{i=1}^{\infty} i = 1 + 2 + 3 + 4 + \cdots, \qquad (4.38)$$

apparently diverges in a normal sense. However, as strange as it seems, it is possible to interpret the summation above from the perspective of analytic continuation and assign a value of $-\frac{1}{12}$ [13]. There are many ways to demonstrate Eq. (4.38) including the statements by Ramanujan [2] and Euler [9].

To illustrate the concept of the sum, consider $f(z)$ defined as

$$f(z) \equiv 1 - z + z^2 - z^3 + \cdots. \qquad (4.39)$$

This geometric series is convergent for $|z| < 1$. On the other hand, another function, $g(z)$, defined as

$$g(z) \equiv \frac{1}{z+1}, \qquad (4.40)$$

exists in the entire plane except for $z = -1$. However, in $|z| < 1$,

$$f(z) \equiv g(z), \tag{4.41}$$

as $f(z)$ is the Taylor series of $g(z)$ about $z = 0$. Hence, $g(z)$ is analytic continuation of $f(z)$. Therefore, while it is not possible to substitute $z = -2$ in $f(z)$, there is no problem in substituting $z = -2$ in $g(z)$. Hence, we can now state

$$1 + 2 + 4 + 8 + 16 + \cdots = -1. \tag{4.42}$$

Following this idea, define the Riemann zeta function, $\zeta(z)$, as

$$\zeta(z) \equiv \frac{1}{1^z} + \frac{1}{2^z} + \frac{1}{3^z} + \frac{1}{4^z} + \cdots = \sum_{n=1}^{\infty} \frac{1}{n^z}. \tag{4.43}$$

If we substitute $z = -1$, Eq. (4.43) becomes Eq. (4.38). However, $\zeta(z)$ is divergent at $z = -1$ and it can be shown that the range of convergence for $\zeta(z)$ is $\Re(z) > 1$ (Example: $\zeta(2) = \frac{\pi^2}{6}$). Nevertheless, according to analytic continuation, $\zeta(z)$ can be extended outside of $\Re(z) > 1$. Let analytic continuation of $\zeta(z)$ be denoted as $\bar{\zeta}(z)$. As $\bar{\zeta}(z)$ is analytic continuation of $\zeta(z)$, for $\Re(z) > 1$, it follows

$$\bar{\zeta}(z) \equiv \zeta(z). \tag{4.44}$$

While $\zeta(z)$ cannot be evaluated for $z = -1$, $\bar{\zeta}(-1)$ can be evaluated using the following functional relationship [1]:

$$\bar{\zeta}(z) = 2^z \pi^{z-1} \Gamma(1-z) \sin\left(\frac{\pi z}{2}\right) \bar{\zeta}(1-z). \tag{4.45}$$

By substituting $z = -1$ on both sides of Eq. (4.45), we have

$$\bar{\zeta}(-1) = -\frac{1}{12}. \tag{4.46}$$

Therefore, we can state

$$1 + 2 + 3 + 4 + 5 + \cdots = -\frac{1}{12}. \tag{4.47}$$

The absolute value of $\zeta(z)$ is shown in Fig. 4.8. It is seen that $|\zeta(z)|$ diverges as $x \to 1$. At $z = -1$, the value of $\zeta(-1)$ takes -0.08333 which is $-\frac{1}{12}$.

4.2   Laurent Series

As discussed in Sect. 4.1, the Taylor series of $f(z)$ around $z = a$ converges within a disk centered at $z = a$ that contains no singularities. A question arises if there is any way that $f(z)$ can be still expanded in a region outside the disk in which the Taylor series fails. It is the Laurent series that takes over where the Taylor series left off to cover $f(z)$ in that region.

In[ ]:= **Plot3D[Abs[Zeta[x + I y]], {x, -3, 3}, {y, -3, 3}, AxesLabel → Automatic]**

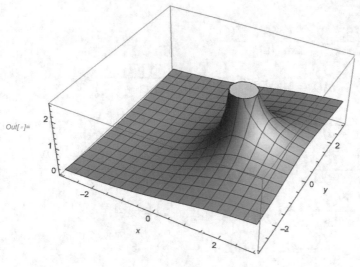

Out[ ]=

In[ ]:= **Zeta[x + I y] /. {x → -1, y → 0}**

Out[ ]= $-\dfrac{1}{12}$

Fig. 4.8 Graph of $|\zeta(z)|$

Using ζ as the active variable, consider Fig. 4.9 in which $\zeta = z_1$ is a singular point and $\zeta = a$ is the center of expansion.

The closed integral path shown in Fig. 4.9 is chosen so as to avoid $\zeta = z_1$. Noting that $f(\zeta)$ is analytic inside the integral path, Cauchy's integral formula at $\zeta = z$ is expressed as

Fig. 4.9 Integral path to derive
the Laurent series

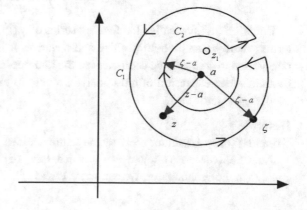

$$f(z) = \frac{1}{2\pi i} \oint_C \frac{f(\zeta)}{\zeta - z} d\zeta$$

$$= \frac{1}{2\pi i} \oint_{C_1} \frac{f(\zeta)}{\zeta - z} d\zeta + \frac{1}{2\pi i} \oint_{-C_2} \frac{f(\zeta)}{\zeta - z} d\zeta$$

$$= \frac{1}{2\pi i} \oint_{C_1} \frac{f(\zeta)}{(\zeta - a)(1 - \frac{z-a}{\zeta - a})} d\zeta + \frac{1}{2\pi i} \oint_{C_2} \frac{f(\zeta)}{(z - a)(1 - \frac{\zeta - a}{z - a})} d\zeta$$

$$= \frac{1}{2\pi i} \oint_{C_1} \frac{f(\zeta)}{(\zeta - a)} \left(1 + \left(\frac{z - a}{\zeta - a} \right) + \left(\frac{z - a}{\zeta - a} \right)^2 + \left(\frac{z - a}{\zeta - a} \right)^3 + \cdots \right) d\zeta$$

$$+ \frac{1}{2\pi i} \oint_{C_2} \frac{f(\zeta)}{(z - a)} \left(1 + \left(\frac{\zeta - a}{z - a} \right) + \left(\frac{\zeta - a}{z - a} \right)^2 + \left(\frac{\zeta - a}{z - a} \right)^3 + \cdots \right) d\zeta$$

$$= \left(\frac{1}{2\pi i} \oint_{C_1} \frac{f(\zeta)}{\zeta - a} d\zeta \right) + \left(\frac{1}{2\pi i} \oint_{C_1} \frac{f(\zeta)}{(\zeta - a)^2} d\zeta \right) (z - a)$$

$$+ \left(\frac{1}{2\pi i} \oint_{C_1} \frac{f(\zeta)}{(\zeta - a)^3} d\zeta \right) (z - a)^2 + \cdots$$

$$+ \left(\frac{1}{2\pi i} \oint_{C_2} f(\zeta) d\zeta \right) \frac{1}{z - a} + \left(\frac{1}{2\pi i} \oint_{C_2} f(\zeta)(\zeta - a) d\zeta \right) \frac{1}{(z - a)^2}$$

$$+ \left(\frac{1}{2\pi i} \oint_{C_2} f(\zeta)(\zeta - a)^2 d\zeta \right) \frac{1}{(z - a)^3} + \cdots$$

$$= c_0 + c_1(z - a) + c_2(z - a)^2 + c_3(z - a)^3 + \cdots$$

$$+ \frac{c_{-1}}{z - a} + \frac{c_{-2}}{(z - a)^2} + \frac{c_{-3}}{(z - a)^3} + \cdots$$

$$= c_0 + \sum_{n=1}^{\infty} c_n(z - a)^n + \sum_{n=1}^{\infty} \frac{c_{-n}}{(z - a)^n}, \tag{4.48}$$

where

$$c_n = \frac{1}{2\pi i} \oint \frac{f(\zeta)}{(\zeta - a)^{n+1}} d\zeta. \tag{4.49}$$

Equation (4.48) is called the Laurent series of $f(z)$ about $z = a$, consisting of $(z - a)^n$ terms where n can be both positive and negative. It is, however, noted that Eq. (4.49) is rarely used to compute the coefficient of the Laurent series. Similar to the Taylor series, the Laurent series expansion of most analytic functions can be carried out from known series such as geometric series.

Example 1

Expand $f(z) = 1/z$ about $z = i$ by the Laurent series.

As the expansion is about $z = i$, it is necessary to isolate $(z - i)$ terms in a manner that $z - i$ is in the denominator. We can write $1/z$ as

Fig. 4.10 Laurent series:
Example 1

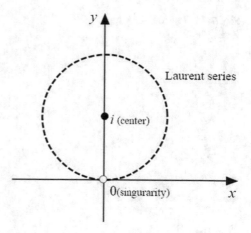

$$\frac{1}{z} = \frac{1}{(z-i)+i}$$

$$= \frac{1}{(z-i)\left(1+\frac{i}{z-i}\right)}$$

$$= \frac{1}{z-i}\left(1 - \left(\frac{i}{z-i}\right) + \left(\frac{i}{z-i}\right)^2 - \left(\frac{i}{z-i}\right)^3 + \cdots \right). \qquad (4.50)$$

Note that we extracted $(z-i)$ rather than i to apply the geometric series as we need negative exponent terms of $(z-i)$. The convergence range of Eq. (4.50) is $|z-i| > 1$ as seen in Fig. 4.10.

Example 2

Expand

$$f(z) = \frac{1}{z+i}, \qquad (4.51)$$

about $z = 0$ for $|z| > 1$ as shown in Fig. 4.11.

The function, $\frac{1}{z+i}$, is singular at $z = -i$. You can draw circles centered at $z = 0$ without hitting a singularity until they reach $z = -i$. Therefore, the region where the Taylor series is valid is $|z| < 1$ and the Laurent series must be used for $|z| > 1$. As $f(z)$ must be expanded by the Laurent series with negative exponents, z must be factored out so that $(1 + i/z)$ is expanded by geometric series as

$$\frac{1}{z+i} = \frac{1}{z(1+\frac{i}{z})}$$

$$= \frac{1}{z}\left(1 - \left(\frac{i}{z}\right) + \left(\frac{i}{z}\right)^2 - \left(\frac{i}{z}\right)^3 + \cdots \right). \qquad (4.52)$$

Fig. 4.11 Laurent series:
Example 2

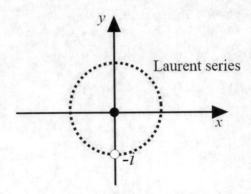

Example 3

Expand

$$f(z) = \frac{1}{z(z-2)},\qquad(4.53)$$

about $z = i$ in the three separate regions shown in Fig. 4.12.

As the center of expansion is $z = i$, circles centered at $z = i$ can be drawn as shown in Fig. 4.12. As the innermost region ($|z + i| < 1$) does not contain any singular point inside, the Taylor series can be used. In the intermediate region ($1 < |z + i| < \sqrt{5}$), the Laurent series must be used for $1/z$ while the Taylor series must be used for $1/(z - 2)$. In the outermost region ($|z + i| > \sqrt{5}$), both $1/z$ and $1/(z - 2)$ must be expanded by the Laurent series.

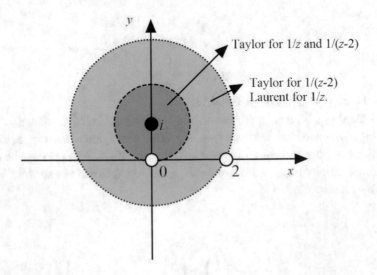

Fig. 4.12 Laurent series: Example 3

(I) $|z - i| < 1$.

As there are no singular points inside the region, both $1/z$ and $1/(z-2)$ must be expanded by the Taylor series as

$$\frac{1}{z} = \frac{1}{(z-i)+i}$$

$$= \frac{1}{i(1 + \frac{z-i}{i})}$$

$$= \frac{1}{i}\left(1 - \left(\frac{z-i}{i}\right) + \left(\frac{z-i}{i}\right)^2 - \left(\frac{z-i}{i}\right)^3 + \cdots\right). \qquad (4.54)$$

$$\frac{1}{z-2} = \frac{1}{(z-i)-(2-i)}$$

$$= \frac{1}{-(2-i)(1 - \frac{z-i}{2-i})}$$

$$= \frac{1}{-(2-i)}\left(1 + \left(\frac{z-i}{2-i}\right) + \left(\frac{z-i}{2-i}\right)^2 + \left(\frac{z-i}{2-i}\right)^3 + \cdots\cdots\right). (4.55)$$

(II) $1 < |z - i| < \sqrt{5}$.

In this region, $1/z$ needs to be expanded by the Laurent series as this accounts for a singular point, $z = 0$, which is inside this region, while $1/(z-2)$ must be expanded by the Taylor series as

$$\frac{1}{z} = \frac{1}{(z-i)+i}$$

$$= \frac{1}{(z-i)(1 + \frac{i}{z-i})}$$

$$= \frac{1}{z-i}\left(1 - \left(\frac{i}{z-i}\right) + \left(\frac{i}{z-i}\right)^2 - \left(\frac{i}{z-i}\right)^3 + \cdots\right). \qquad (4.56)$$

$$\frac{1}{z-2} : \text{same as Eq. (4.55).}$$

(III) $|z| > \sqrt{5}$.

This time, $1/(z-2)$ needs to be expanded by the Laurent series as

Fig. 4.13 The area where $\frac{1}{\sin z}$ is to be expanded by the Laurent series

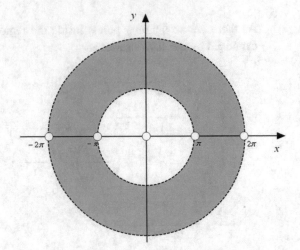

$$
\begin{aligned}
\frac{1}{z-2} &= \frac{1}{(z-i)-(2-i)} \\
&= \frac{1}{(z-i)(1-\frac{2-i}{z-i})} \\
&= \frac{1}{z-i}\left(1+\left(\frac{2-i}{z-i}\right)+\left(\frac{2-i}{z-i}\right)^2+\left(\frac{2-i}{z-i}\right)^3+\cdots\right).
\end{aligned}
\tag{4.57}
$$

Example 4

Expand

$$
f(z) = \frac{1}{\sin z},
\tag{4.58}
$$

about $z = 0$ over $\pi < |z| < 2\pi$ shown in Fig. 4.13.

Note that the singularities of $1/\sin z$ are distributed on the real axis as

$$
z = 0, \pm\pi, \pm 2\pi, \pm 3\pi, \ldots
\tag{4.59}
$$

The region, $\pi < |z| < 2\pi$, shown in Fig. 4.13 contains three singular points, i.e., $z = 0, \pm\pi$. At each singular point, $\sin z$ can be expanded as

$$
\text{At } z = 0, \quad \sin z = z - \frac{z^3}{3!} + \frac{z^5}{5!} - \cdots,
\tag{4.60}
$$

$$\text{At } z = \pi, \quad \sin z = \sin((z - \pi) + \pi)$$

$$= -\sin(z - \pi)$$

$$= -\left((z - \pi) - \frac{(z - \pi)^3}{3!} + \frac{(z - \pi)^5}{5!} - \cdots\right), \quad (4.61)$$

$$\text{At } z = -\pi, \quad \sin z = \sin((z + \pi) - \pi)$$

$$= -\sin(z + \pi)$$

$$= -\left((z + \pi) - \frac{(z + \pi)^3}{3!} + \frac{(z + \pi)^5}{5!} - \cdots\right). \quad (4.62)$$

Therefore, $\sin z$ can be expanded at each singular point asymptotically[3] as

$$\sin z \sim z \qquad \text{as } z \to 0, \quad (4.63)$$

$$\sin z \sim -(z - \pi) \quad \text{as } z \to \pi, \quad (4.64)$$

$$\sin z \sim -(z + \pi) \quad \text{as } z \to -\pi. \quad (4.65)$$

It follows

$$\frac{1}{\sin z} = \frac{z(z - \pi)(z + \pi)}{\sin z} \frac{1}{z(z - \pi)(z + \pi)}$$

$$= g(z) \frac{1}{z(z - \pi)(z + \pi)}, \quad (4.66)$$

where

$$g(z) \equiv \frac{z(z - \pi)(z + \pi)}{\sin z} \quad (4.67)$$

is analytic in the region $\pi < |z| < 2\pi$ which can be expanded by the Taylor series as

$$g(z) = g(0) + g'(0)z + \frac{g''(0)}{2!}z^2 + \frac{g'''(0)}{3!}z^3 + \cdots. \quad (4.68)$$

It is straightforward to expand

$$\frac{1}{z(z - \pi)(z + \pi)}, \quad (4.69)$$

by the Laurent series as

---

[3] The statement, "$f \sim g$ as $x \to x_0$," means $\lim_{x \to x_0} \frac{f(x)}{g(x)} = 1$.

$$\frac{1}{z(z-\pi)(z+\pi)} = \frac{1}{z}\frac{1}{z\left(1-\frac{\pi}{z}\right)}\frac{1}{z\left(1+\frac{\pi}{z}\right)}$$

$$= \frac{1}{z^3}\left(1+\left(\frac{\pi}{z}\right)+\left(\frac{\pi}{z}\right)^2+\left(\frac{\pi}{z}\right)^3+\cdots\right)\times$$

$$\left(1-\left(\frac{\pi}{z}\right)+\left(\frac{\pi}{z}\right)^2-\left(\frac{\pi}{z}\right)^3+\cdots\right). \qquad (4.70)$$

Finally, combining all, we have

$$\frac{1}{\sin z} = \left(-\pi^2+\left(1-\frac{\pi^2}{6}\right)z^2+\left(\frac{1}{6}-\frac{7\pi^2}{360}\right)z^4+\cdots\right)\times$$

$$\frac{1}{z^3}\left(1+\left(\frac{\pi}{z}\right)+\left(\frac{\pi}{z}\right)^2+\left(\frac{\pi}{z}\right)^3+\cdots\right)\times$$

$$\left(1-\left(\frac{\pi}{z}\right)+\left(\frac{\pi}{z}\right)^2-\left(\frac{\pi}{z}\right)^3+\cdots\right). \qquad (4.71)$$

The idea employed here is to modify $f(z)$ to be

$$g(z)\times\frac{1}{z(z-\pi)(z+\pi)}, \qquad (4.72)$$

where $g(z)$ is analytic everywhere on and inside $0 < |z| < 2\pi$. As $g(z)$ has no singular point inside $0 < |z| < 2\pi$, it can be expanded by the Taylor series. All the singularities were passed to $\frac{1}{z(z-\pi)(z+\pi)}$ which is easy to expand by the Laurent series.

4.3   *Mathematica Code*

The function, Series[f[z],{z, a, n}], returns the Taylor series of f[z]. However, there is no built-in function that performs the Laurent series expansion in *Mathematica*. A workaround is to use the Series[] with $z = \infty$ as the center of expansion. For example, the Laurent series of $f(z) = \frac{1}{z-1}$ about $z = 0$ up to the tenth order is computed by

```
In[·]:= Series[1/(z-1), {z, Infinity, 10}]
```

$$Out[·]= \frac{1}{z}+\left(\frac{1}{z}\right)^2+\left(\frac{1}{z}\right)^3+\left(\frac{1}{z}\right)^4+\left(\frac{1}{z}\right)^5+\left(\frac{1}{z}\right)^6+\left(\frac{1}{z}\right)^7+\left(\frac{1}{z}\right)^8+\left(\frac{1}{z}\right)^9+\left(\frac{1}{z}\right)^{10}+O\left[\frac{1}{z}\right]^{11}$$

If the center of the expansion in the Laurent series is other than $z = 0$, it is necessary to shift z by a. The following code demonstrates the Laurent series of $f(z) = \frac{1}{z-1}$ about $z = i$ up to the tenth order.

In[ ]:= `Normal[Series[1/(z-1) /. z → w + I, {w, Infinity, 10}]] /. w → z - I`

Out[ ]= $\dfrac{16-16\,i}{(-i+z)^{10}} + \dfrac{16}{(-i+z)^9} + \dfrac{8+8\,i}{(-i+z)^8} + \dfrac{8\,i}{(-i+z)^7} -$

$\dfrac{4-4\,i}{(-i+z)^6} - \dfrac{4}{(-i+z)^5} - \dfrac{2+2\,i}{(-i+z)^4} - \dfrac{2\,i}{(-i+z)^3} + \dfrac{1-i}{(-i+z)^2} + \dfrac{1}{-i+z}$

Example 4 above can be computed with the following code:

In[ ]:= `t1 = Series[z (z - Pi) (z + Pi) / Sin[z], {z, 0, 10}]`

Out[ ]= $-\pi^2 + \left(1 - \dfrac{\pi^2}{6}\right) z^2 + \left(\dfrac{1}{6} - \dfrac{7\,\pi^2}{360}\right) z^4 + \left(\dfrac{7}{360} - \dfrac{31\,\pi^2}{15120}\right) z^6 +$

$\left(\dfrac{31}{15120} - \dfrac{127\,\pi^2}{604800}\right) z^8 + \left(\dfrac{127}{604800} - \dfrac{73\,\pi^2}{3421440}\right) z^{10} + O[z]^{11}$

In[ ]:= `t2 = Series[1/((z - Pi) (z + Pi) z), {z, Infinity, 10}]`

Out[ ]= $\left(\dfrac{1}{z}\right)^3 + \dfrac{\pi^2}{z^5} + \dfrac{\pi^4}{z^7} + \dfrac{\pi^6}{z^9} + O\left[\dfrac{1}{z}\right]^{11}$

In[ ]:= `t1 t2`

... Series: Series in z to be combined have unequal expansion points 0 and ∞.

Out[ ]= $\left(-\pi^2 + \left(1 - \dfrac{\pi^2}{6}\right) z^2 + \left(\dfrac{1}{6} - \dfrac{7\,\pi^2}{360}\right) z^4 + \left(\dfrac{7}{360} - \dfrac{31\,\pi^2}{15120}\right) z^6 + \left(\dfrac{31}{15120} - \dfrac{127\,\pi^2}{604800}\right) z^8 +\right.$

$\left.\left(\dfrac{127}{604800} - \dfrac{73\,\pi^2}{3421440}\right) z^{10} + O[z]^{11}\right) \left(\left(\dfrac{1}{z}\right)^3 + \dfrac{\pi^2}{z^5} + \dfrac{\pi^4}{z^7} + \dfrac{\pi^6}{z^9} + O\left[\dfrac{1}{z}\right]^{11}\right)$

4.4 Problems

1. Simplify

$$1 + \cos\theta + \cos 2\theta + \cdots + \cos(n-1)\theta. \tag{4.73}$$

2. Obtain the first three non-zero terms of the Taylor series for

$$f(z) = \frac{z^2}{1 - \cos z} \quad \text{for} \quad |z| < 2\pi, \tag{4.74}$$

using long division.

3. Explain

$$1 + 2 + 4 + 8 + 16 + \cdots = -1, \tag{4.75}$$

using analytic continuation. Hint: $f(z) = \frac{1}{1-2z}$.

4. Prove that

$$\sin(2(x + iy)) = 2\sin(x + iy)\cos(x + iy), \tag{4.76}$$

where

$$\sin z = \frac{1}{2i}(e^{iz} - e^{-iz}), \quad \cos z = \frac{1}{2}(e^{iz} + e^{-iz}). \tag{4.77}$$

5. Obtain the first non-zero three terms of the Laurent series for

$$f(z) = \frac{1}{e^z - 1} \quad \text{for} \quad |z| > 0. \tag{4.78}$$

Residues

5

The residue of $f(z)$ at $z = a$ is the coefficient of $\frac{1}{z-a}$ in the Laurent series expansion about $z = a$. It is called a residue because when $f(z)$ is integrated along a closed path containing $z = a$, the value of the integration is expressed in terms of the residue at $z = a$. The major application of residues is found in evaluating certain types of improper integrals and integrals containing trigonometric functions.

5.1 Types of Singularities

If the Laurent series expansion of $f(z)$ in the *immediate* neighborhood of $z = a$ is expressed as

$$f(z) = \frac{c_{-m}}{(z-a)^m} + \frac{c_{-m+1}}{(z-a)^{m-1}} + \cdots + \frac{c_{-1}}{z-a} + c_0 + c_1(z-a) + \cdots, \qquad (5.1)$$

$z = a$ is called an m-th order pole or m-th order singularity. If $m = \infty$, $z = a$ is called an essential singularity. Otherwise, singular points are called *removable singularities*.

Example 1

Identify the order of singularity at $z = 0$ for

$$f(z) = \frac{1}{z^2(1-z)}. \qquad (5.2)$$

© The Author(s), under exclusive license to Springer Nature Switzerland AG 2022
S. Nomura, *Complex Variables for Engineers with Mathematica*,
Synthesis Lectures on Mechanical Engineering,
https://doi.org/10.1007/978-3-031-13067-0_5

By expanding $1/(1-z)$ by the Taylor series about $z = 0$, we have

$$f(z) = \frac{1}{z^2}(1 + z + z^2 + z^3 + \cdots)$$

$$= \frac{1}{z^2} + \frac{1}{z} + 1 + z + z^2 + z^3 + \cdots . \tag{5.3}$$

Therefore, $z = 0$ is a second order pole.

The Series[] function in *Mathematica* can be used for the Taylor/Laurent series.

$$\textit{In[\cdot]:=} \ \textbf{Series}[\textbf{1/}(\textbf{z\textasciicircum2 (1-z)), \{z, 0, 10\}}]$$

$$\textit{Out[\cdot]=} \ \frac{1}{z^2} + \frac{1}{z} + 1 + z + z^2 + z^3 + z^4 + z^5 + z^6 + z^7 + z^8 + z^9 + z^{10} + O[z]^{11}$$

Example 2

Identify the type of singularities for

$$f(z) = \frac{1}{z^2} + \frac{1}{z^3} + \frac{1}{z^4} + \frac{1}{z^5} + \cdots = \sum_{n=1}^{\infty} \frac{1}{z^{n+1}}. \tag{5.4}$$

It first appears that $z = 0$ is an essential singularity as $m = \infty$. However, from the ratio test, the range of convergence for $f(z)$ is

$$\lim_{n \to \infty} \left| \frac{a_{n+1}}{a_n} \right| = \lim_{n \to \infty} \left| \frac{z^{n+1}}{z^{n+2}} \right| = \left| \frac{1}{z} \right| < 1, \tag{5.5}$$

i.e.,

$$|z| > 1, \tag{5.6}$$

which is not an immediate neighborhood of $z = 0$. On the other hand, by factoring out $1/z^2$, $f(z)$ can be written as

$$f(z) = \frac{1}{z^2}\left(1 + \left(\frac{1}{z}\right) + \left(\frac{1}{z}\right)^2 + \left(\frac{1}{z}\right)^3 + \cdots\right)$$

$$= \frac{1}{z^2} \frac{1}{1 - \frac{1}{z}}$$

$$= \frac{1}{z(z-1)}$$

$$= -\frac{1}{z} - 1 - z - z^2 - z^3 - \cdots , \tag{5.7}$$

so $z = 0$ is actually a first order pole. To check the order of singularity, the function must be expanded in the *immediate* neighborhood of the singular point.

The procedure above can be implemented in *Mathematica* with the following code:

```
In[ ]:= Sum[1/z^i, {i, 2, Infinity}]
```

$$Out[ ]= \frac{1}{(-1+z)\,z}$$

```
In[ ]:= Series[%, {z, 0, 5}]
```

$$Out[ ]= -\frac{1}{z} - 1 - z - z^2 - z^3 - z^4 - z^5 + O[z]^6$$

The "%" symbol refers to the preceding output.

Example 3
Identify the type of singularities at $z = 0$ for

$$f(z) = e^{-\frac{1}{z^2}}. \tag{5.8}$$

Expanding $f(z)$ about $z = 0$ yields

$$e^{-\frac{1}{z^2}} = 1 - \frac{1}{z^2} + \frac{1}{2!}\frac{1}{z^4} - \frac{1}{3!}\frac{1}{z^6} + \cdots, \tag{5.9}$$

This time, $z = 0$ is indeed an essential singular point. At an essential singular point, the function behaves rather wildly. To see how this function behaves, let z approach 0 along the real axis. By setting $z = \epsilon$, we have

$$f(z) \to \frac{1}{e^{1/\epsilon^2}} \to \frac{1}{e^{(\text{a very large number})}} \to \frac{1}{\text{an even larger number}} \to 0. \tag{5.10}$$

On the other hand, if z approaches 0 along the imaginary axis, z can be set as $z = \epsilon i$. It follows

$$f(z) \to \frac{1}{e^{1/(i^2\epsilon^2)}} \to e^{(\text{a very large number})} \to (\text{an even larger number}) \to \infty. \tag{5.11}$$

If z approaches 0 along the line, $y = x$, z can be set as $z = \epsilon(1 + i)$ and

$$f(z) \to e^{-\frac{1}{(1+i)^2\epsilon^2}}$$
$$= e^{-\frac{1}{2\epsilon^2 i}}$$
$$= e^{\frac{i}{2\epsilon^2}}$$
$$= \cos\frac{1}{2\epsilon^2} + i\sin\frac{1}{2\epsilon^2}, \tag{5.12}$$

which fluctuates perpetually. It is not possible to draw the graph of $|e^{-1/z^2}|$, but we can still see its wild behavior in Fig. 5.1.

```
In[*]:= z = x + I y;
        Plot3D[ Abs[Exp[-1 / z^2]], {x, -0.1, 0.1},
           {y, -0.1, 0.1}, PlotRange → {0, 100000}, AxesLabel → Automatic]
```

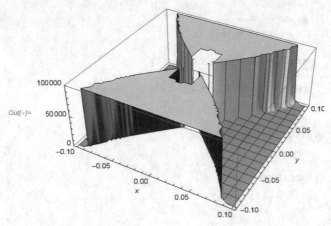

Fig. 5.1 Forced 3-D plot of $e^{-\frac{1}{z^2}}$

According to Great Picard's theorem [4], if an analytic function $f(z)$ has an essential singularity at a point a, then $f(z)$ can take all possible complex values about $z = a$.

5.2 Residues

5.2.1 Definition of Residues

When $f(z)$ is expanded by the Laurent series about an m-th order pole at $z = a$ as

$$f(z) = \frac{c_{-m}}{(z-a)^m} + \frac{c_{-m+1}}{(z-a)^{m-1}} + \cdots + \frac{c_{-1}}{(z-a)} + c_0 + c_1(z-a) + \cdots , \qquad (5.13)$$

the coefficient, c_{-1}, of $1/(z-a)$ is called the *residue* at that singular point and is denoted as

$$c_{-1} = \text{Res}\,(f(z); a) . \qquad (5.14)$$

5.2.2 Calculation of Residues

It is not effective to compute the residue of $f(z)$ by first expanding $f(z)$ with the Laurent series and by extracting the coefficient of $1/(z-a)$ as we do not require the rest of the coefficients. A better way to compute the residues is available that bypasses the Laurent series expansion.

First Order Poles

If $z = a$ is a first order pole, the Laurent series of $f(z)$ is expressed as

$$f(z) = \frac{c_{-1}}{(z-a)} + c_0 + c_1(z-a) + c_2(z-a)^2 + \cdots . \tag{5.15}$$

In order to extract c_{-1}, we multiply $(z-a)$ on both sides of Eq. (5.15) to obtain

$$(z-a)f(z) = c_{-1} + c_0(z-a) + c_1(z-a)^2 + c_2(z-a)^3 + \cdots . \tag{5.16}$$

Therefore,

$$c_{-1} = \lim_{z \to a}(z-a)f(z). \tag{5.17}$$

- Example: Compute the residue at $z = 1$ for

$$f(z) = \frac{1}{z(z-1)}. \tag{5.18}$$

As $z = 1$ is a first order pole, it follows

$$c_{-1} = \lim_{z \to 1}(z-1)f(z) = \lim_{z \to 1}\frac{1}{z} = 1. \tag{5.19}$$

Second Order Poles

If $z = a$ is a second order pole, the Laurent series of $f(z)$ is expressed as

$$f(z) = \frac{c_{-2}}{(z-a)^2} + \frac{c_{-1}}{(z-a)} + c_0 + c_1(z-a) + c_2(z-a)^2 + \cdots . \tag{5.20}$$

By multiplying $(z-a)^2$ on both sides of Eq. (5.20), we obtain

$$(z-a)^2 f(z) = c_{-2} + c_{-1}(z-a) + c_0(z-a)^2 + c_1(z-a)^3 + \cdots . \tag{5.21}$$

If we substitute $z = a$ in Eq. (5.21), we will pick up c_{-2} instead of c_{-1}. However, noting that Eq. (5.21) is an identity, we can obtain yet another identity by differentiating both sides of Eq. (5.21) as

$$\frac{d}{dz}\left((z-a)^2 f(z)\right) = c_{-1} + 2c_0(z-a) + 3c_1(z-a)^2 + \cdots . \tag{5.22}$$

Therefore,

$$c_{-1} = \lim_{z \to a} \left\{ \frac{d}{dz} \left((z-a)^2 f(z) \right) \right\}.$$ (5.23)

Higher Order Poles

Following the method above, if $z = a$ is an m-th order pole, the residue of $f(z)$ at $z = a$ can be obtained as

$$c_{-1} = \frac{1}{(m-1)!} \lim_{z \to a} \left\{ \frac{d^{m-1}}{dz^{m-1}} \left((z-a)^m f(z) \right) \right\}.$$ (5.24)

Shortcut (Residues for First Order Poles Only)

If $z = a$ is a first order pole, there is an alternative (and faster) way to calculate the residue. Let $f(z)$ be expressed as

$$f(z) = \frac{h(z)}{g(z)},$$ (5.25)

then

$$c_{-1} = \frac{h(a)}{g'(a)}.$$ (5.26)

Proof

$$\begin{aligned} c_{-1} &= \lim_{z \to a} (z-a) f(z) \\ &= \lim_{z \to a} \frac{(z-a)h(z)}{g(z)} \\ &= \lim_{z \to a} \frac{h(z) + (z-a)h'(z)}{g'(z)} \\ &= \frac{h(a)}{g'(a)}, \end{aligned}$$ (5.27)

where L'Hôspital's rule[1] was used from the second line to the third line in Eq. (5.27).

Example

Find all the residues for

$$f(z) = \frac{1}{z^4 + 1}.$$ (5.29)

---

[1] If $f(a) = g(a) = 0$, it follows

$$\lim_{x \to a} \frac{f(x)}{g(x)} = \frac{f(a)}{g(a)}.$$ (5.28)

There are four zeros for $z^4 + 1 = 0$ each of which is a first order pole. The poles are[2]

$$z = e^{\pi i/4}, \quad e^{3\pi i/4}, \quad e^{5\pi i/4}, \quad e^{7\pi i/4}. \tag{5.33}$$

For example, the residue at $z = e^{\pi i/4}$ is

$$\begin{aligned}
\text{Res}\left(f(z); e^{\pi i/4}\right) &= \frac{1}{(z^4 + 1)'|_{z \to e^{\pi i/4}}} \\
&= \frac{1}{4z^3|_{z \to e^{\pi i/4}}} \\
&= \frac{1}{4e^{3\pi i/4}} \\
&= \frac{1}{4}e^{-3\pi i/4} \\
&= -\frac{1}{4}\left(\frac{1}{\sqrt{2}} + \frac{1}{\sqrt{2}}i\right).
\end{aligned} \tag{5.34}$$

Mathematica **Code**

In *Mathematica*, you can use a built-in function, Residue[f[z], {z, a}], to compute the residue of f[z] at z = a as

```
In[ ]:= f[z_] := 1 / (z^4 + 1)

In[ ]:= Residue[f[z], {z, Exp[Pi I / 4]}]

Out[ ]=  -1/4 (-1)^(1/4)

In[ ]:= ComplexExpand[%]

Out[ ]=  -( (1/4 + i/4) / √2 )
```

---

[2] The singularities are the zeros to the equation,

$$z^4 + 1 = 0 \quad \text{or} \quad z^4 = -1. \tag{5.30}$$

Expressing both z and -1 in polar form, the equation above becomes

$$r^4 e^{4i\theta} = 1e^{(\pi + 2n\pi)i} \tag{5.31}$$

from which it follows

$$r = 1, \quad \theta = \frac{\pi}{4} + \frac{n\pi}{2}, \tag{5.32}$$

where $n = 0, 1, 2, 3$.

The `ComplexExpand[]` function expands a complex number into the real part and the imaginary part assuming all the variables are real.

5.3   Residue Theorem

Theorem 7 *If $f(z)$ has isolated singular points, $z_1, z_2, z_3, \ldots, z_n$, inside a closed loop, C, as shown in Fig. 5.2, it follows*

$$\oint_C f(z)dz = 2\pi i \sum_{i=1}^{n} Res(f; z_i), \tag{5.35}$$

where the integral path, C, is a CCW closed loop that sees all the poles on the left.

Proof The integration over the path, C, can be altered without changing any topological relationships to an integration along the deformed path shown in the middle figure of Fig. 5.3, which is further split into isolated integrations over the loops each of which encircles each of the poles as shown in the right figure of Fig. 5.3 because the integrations along both sides of the connecting paths are 0 due to the opposite signs. Therefore,

$$\oint_C f(z)dz = \sum_{i=1}^{n} \oint_{C_i} f(z)dz$$

$$= \sum_{i=1}^{n} \oint_{C_i} \left(\cdots + \frac{c_{-1}}{(z - z_i)} + c_0 + c_1(z - z_i) + \cdots \right) dz$$

$$= 2\pi i \sum_{i=1}^{n} Res(f; z_i), \tag{5.36}$$

Fig. 5.2 A contour that contains multiple singular points

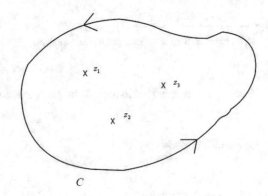

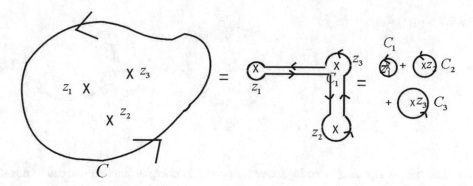

Fig. 5.3 Deformation and separation of the original contour

where

$$\oint_{C_i} \frac{1}{z - z_i} dz = 2\pi i \tag{5.37}$$

was used. □

5.3.1 Residue at Infinity

It is possible to examine poles at $z = \infty$ by considering the Riemann sphere shown in Fig. 5.4 where a sphere is placed on the complex plane and the bottom of the sphere coincides with the origin, $z = 0$. A straight line connecting the North Pole of the sphere, N, and a point, P, on the surface of the sphere is projected to a point, z, on the complex plane. This enables a one-to-one correspondence between P and z, and all the points on the complex plane have their equivalence on the Riemann sphere. However, this mapping fails when P is approaching N as the straight line from N will not intersect with the complex plane. Nevertheless, we can define $z = \infty$ as a point on the complex plane that corresponds to the North Pole, N, on the

Fig. 5.4 Riemann sphere

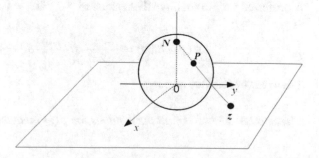

Fig. 5.5 C_0 and C_∞

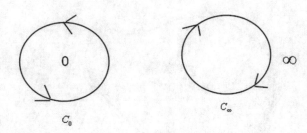

$$C_0 \qquad\qquad\qquad C_\infty$$

Riemann sphere. Hence, $z = \infty$ is just another ordinary complex number and can be added to the set of complex numbers for completeness.

The Laurent series of $f(z)$ about $z = \infty$ can be expressed by the equivalent Laurent series of $g(\zeta) \equiv f(\frac{1}{z})$ about $\zeta = 0$ as

$$g(\zeta) = \sum_{n=-\infty}^{\infty} c_n \zeta^n, \quad c_n = \frac{1}{2\pi i} \oint_{C_0} \frac{g(\zeta)}{\zeta^{n+1}} d\zeta, \tag{5.38}$$

where C_0 is a counter-clockwise closed path which sees $\zeta = 0$ on the left as shown in the left figure of Fig. 5.5. Rewriting c_n in Eq. (5.38) in terms of z yields

$$c_n = \frac{1}{2\pi i} \oint_{C_\infty} \frac{f(z)}{(\frac{1}{z})^{n+1}} \frac{d\zeta}{dz} dz = -\frac{1}{2\pi i} \oint_{C_\infty} \frac{f(z)}{z^{1-n}} dz, \tag{5.39}$$

where C_∞ is a clockwise path that sees $z = \infty$ on the left as shown in the right figure of Fig. 5.5.

The residue at $z = \infty$ is defined as

$$\text{Res}\,(f(z); \infty) \equiv \frac{1}{2\pi i} \oint_{C_\infty} f(z) dz. \tag{5.40}$$

Comparing Eq. (5.40) with Eq. (5.39), we have

$$\text{Res}\,(f(z); \infty) = -c_1. \tag{5.41}$$

A preferred way to compute the residue of $f(z)$ at $z = \infty$ is by setting $n = 1$ in Eq. (5.38) as

$$\text{Res}\,(f(z); \infty) = -\frac{1}{2\pi i} \oint_{C_0} \frac{g(\zeta)}{\zeta^2} d\zeta = \text{Res}\left(\frac{f(\frac{1}{\zeta})}{\zeta^2}; 0 \right). \tag{5.42}$$

Theorem of Residue at ∞

Theorem 8 *The sum of all the residues for $f(z)$ including $z = \infty$ is 0.*

Proof Consider a circle $|z| = R$ on the complex plane where R is sufficiently large so that this circle encloses all the poles of $f(z)$ except for $z = \infty$.[3] By reverting the direction of the integral path in Eq. (5.40), we have

$$\text{Res}\,(f(z); \infty) = -\frac{1}{2\pi i} \oint_{|z|=R} f(z) dz, \tag{5.43}$$

where the integral path is such that the origin is seen on the left. On the other hand, from the Residue theorem, we have

$$\sum_i \text{Res}\,(f(z); \alpha_i) = \frac{1}{2\pi i} \oint_{|z|=R} f(z) dz. \tag{5.44}$$

Therefore,

$$\sum_i \text{Res}\,(f(z); \alpha_i) + \text{Res}\,(f(z); \infty) = 0. \tag{5.45}$$

$\square$

This theorem is useful when there are many poles within the integral path, C, as you only need to compute the residue at $z = \infty$.

Example
Integrate

$$f(z) = \frac{17}{z-2} + \frac{\pi}{z} + \frac{9}{z-i}, \tag{5.46}$$

along $|z| = 10$ counter-clock wise.

The poles are at $z = 2$, $z = 0$ and $z = i$. As $|z| = 10$ contains all the poles, following the residue theorem, we have

$$\frac{1}{2\pi i} \oint_{|z|=10} f(z) dz = \text{Res}(f; 2) + \text{Res}(f; 0) + \text{Res}(f; i)$$
$$= 17 + \pi + 9$$
$$= 26 + \pi. \tag{5.47}$$

However, instead of computing all the residues and summing them up, it is faster to compute the residue at $z = \infty$ and change the sign by taking advantage of Theorem 8. We have

---

[3] Note that for $z = \infty$ to be a pole, $f(z)$ must be analytic in $|z| > R$. For example, $f(z) = \sin(z)$ has poles, $z = \pm n\pi$, on the real axis, and there is no R such that $f(z)$ is analytic for $|z| > R$, hence, $z = \infty$ is not a pole.

$$\frac{f(\frac{1}{z})}{z^2} = \frac{2\pi z^2 - (1-2i)\pi z - (17-18i)z - i\pi - 26i}{z(z+i)(2z-1)}$$

$$= 281z^3 + (136-9i)z^2 + 59z + \frac{26+\pi}{z} + (34+9i) + \cdots . \qquad (5.48)$$

Therefore, the residue of $f(z)$ at $z = \infty$ is

$$\mathrm{Res}\,(f(z); \infty) = -\mathrm{Res}\left(\frac{f(\frac{1}{z})}{z^2}; 0\right) = -(26+\pi), \qquad (5.49)$$

which matches Eq. (5.47).

Mathematica can compute the residue at infinity using the built-in constant, `Infinity`, as

```
In[*]:= f[z_] := 17 / (z - 2) + Pi / z + 9 / (z - I)

In[*]:= Residue[f[z], {z, Infinity}]

Out[*]= - 26 - π
```

5.4   Application of Residue Theorem to Certain Integrals

One of the applications of the residue theorem is the evaluation of certain types of real integrals. If a given integral happens to belong to one of the three types below, its evaluation is done routinely by hand. The three types of integrals where the residue theorem can be used are

1.
$$\int_0^{2\pi} f(\sin\theta, \cos\theta)d\theta. \qquad (5.50)$$

   Examples:
$$\int_0^{2\pi} \frac{d\theta}{2+3\cos\theta}, \quad \int_0^{2\pi} \cos^4\theta d\theta. \qquad (5.51)$$

2.
$$\int_{-\infty}^{\infty} f(x)dx \quad \text{or} \quad \int_{-\infty}^{\infty} f(x)e^{ix}dx. \qquad (5.52)$$

   Examples:
$$\int_{-\infty}^{\infty} \frac{x^2}{1+x^6}dx, \quad \int_{-\infty}^{\infty} \frac{x\sin x}{1+x^2}dx. \qquad (5.53)$$

3.
$$\int_0^{\infty} f(x^\alpha, \log x)dx, \quad \alpha \text{ is non-integer}. \qquad (5.54)$$

Examples:

$$\int_0^\infty \frac{1}{\sqrt{x}(1+x)} dx, \quad \int_0^\infty \frac{(\log x)^2}{1+x^2} dx. \tag{5.55}$$

5.4.1 First Type

The first type of integrals, Eq. (5.50), must satisfy the following conditions:

1. The function, f, must be a function of $\cos \theta$ or $\sin \theta$ or their variations.
2. The integral range must be from 0 to 2π.

An integral belonging to the first type can be converted into an equivalent complex integral along the unit circle, thus, the residue theorem can be used.

The first step is to change the variable from θ to z by setting

$$z = e^{i\theta}. \tag{5.56}$$

The integral from 0 to 2π with respect to θ is converted into a contour integral along the unit circle as

$$\int_0^{2\pi} \rightarrow \oint_{|z|=1}. \tag{5.57}$$

Taking total derivative of Eq. (5.56) yields the relationship between dz and $d\theta$ as

$$\begin{aligned} dz &= i \, e^{i\theta} \, d\theta \\ &= i \, z \, d\theta, \end{aligned} \tag{5.58}$$

or

$$d\theta = \frac{1}{iz} dz. \tag{5.59}$$

The trigonometric functions can be converted into functions of z as

$$\begin{aligned} \sin \theta &= \frac{1}{2i} \left(e^{i\theta} - e^{-i\theta} \right) \\ &= \frac{1}{2i} \left(z - \frac{1}{z} \right), \end{aligned} \tag{5.60}$$

and

$$\begin{aligned} \cos \theta &= \frac{1}{2} \left(e^{i\theta} + e^{-i\theta} \right) \\ &= \frac{1}{2} \left(z + \frac{1}{z} \right). \end{aligned} \tag{5.61}$$

Therefore, the original integral is now expressed as a complex integral along the unit circle as

$$\int_0^{2\pi} f(\sin\theta, \cos\theta) d\theta = \oint_{|z|=1} f(z) \frac{dz}{iz}$$

$$= 2\pi i \sum_j \text{Res}\left(\frac{f(z)}{iz}; a_j\right), \qquad (5.62)$$

where a_js are the poles inside the unit circle only.

Example 1

Evaluate

$$I \equiv \int_0^{2\pi} \frac{d\theta}{3 + \sin\theta}. \qquad (5.63)$$

The real integral with respect to θ can be converted to a complex integral as

$$\int_0^{2\pi} \frac{d\theta}{3 + \sin\theta} = \oint_{|z|=1} \frac{\frac{dz}{iz}}{3 + \frac{1}{2i}\left(z - \frac{1}{z}\right)}$$

$$= 2 \oint_{|z|=1} \frac{1}{z^2 + 6iz - 1} dz$$

$$= 4\pi i \sum_i \text{Res}\left(\frac{1}{z^2 + 6iz - 1}; a_i\right). \qquad (5.64)$$

The roots of the equation, $z^2 + 6iz - 1 = 0$, are $\left(-3 \pm 2\sqrt{2}\right)i$ of which only $\left(-3 + 2\sqrt{2}\right)i$ is inside the unit circle. Therefore,

$$4\pi i \, \text{Res}\left(\frac{1}{z^2 + 6iz - 1}; \left(-3 + 2\sqrt{2}\right)i\right) = 4\pi i \frac{1}{2z + 6i}\Big|_{z=\left(-3+2\sqrt{2}\right)i}$$

$$= \frac{\pi}{\sqrt{2}}. \qquad (5.65)$$

Note that we used Eq. (5.27) to compute the residue at the first order pole.

Example 2

Evaluate

$$I \equiv \int_0^{2\pi} \cos^6\theta d\theta. \qquad (5.66)$$

It can be shown that $z = 0$ is a seventh order pole after converting the integral in terms of z. However, instead of using Eq. (5.24) to compute the residue at the seventh order pole, it is more straightforward to single out the $1/z$ term out of the expansion of the function of z as

$$\int_0^{2\pi} \cos^6 \theta d\theta = \oint_{|z|=1} \left(\frac{1}{2} \left(z + \frac{1}{z} \right) \right)^6 \frac{dz}{iz}$$

$$= \frac{1}{64i} \oint_{|z|=1} \frac{(1+z^2)^6}{z^7} dz$$

$$= \frac{1}{64i} \oint_{|z|=1} \frac{z^{12} + 6z^{10} + 15z^8 + 20z^6 + 15z^4 + 6z^2 + 1}{z^7} dz$$

$$= \frac{1}{64i} \left(\cdots + \oint_{|z|=1} 20 \frac{dz}{z} + \cdots \right)$$

$$= \frac{1}{64i} \times 20 \times 2\pi i$$

$$= \frac{5}{8} \pi. \tag{5.67}$$

5.4.2 Second Type

The second type of integrals that can be evaluated using the residue theorem is an improper integral in the format of

$$\int_{-\infty}^{\infty} \frac{P(x)}{Q(x)} dx, \tag{5.68}$$

where both $P(x)$ and $Q(x)$ are polynomials on x. For this method to work, the following two conditions must be met:

1. The degree of $Q(x)$ must be larger than the degree of $P(x)$ by at least 2 as

$$\deg \frac{Q(x)}{P(x)} \geq 2. \tag{5.69}$$

2. There is no real x such that $Q(x) = 0$.[4]

When the two conditions above are satisfied, the integral can be carried out as

$$\int_{-\infty}^{\infty} \frac{P(x)}{Q(x)} dx = 2\pi i \sum_{\text{(Upper half plane)}} \text{Res} \left(\frac{P(z)}{Q(z)}; a_i \right), \tag{5.70}$$

where a_i's are poles in the upper half-plane.

---

[4] This will exclude such integrals as $\int_{-\infty}^{\infty} \frac{e^{ix}}{x} dx$.

Fig. 5.6 Singular points in the
upper half-plane

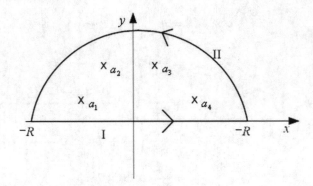

Proof Referring to Fig. 5.6, from the Residue theorem, we have

$$\oint_{I+II} f(z)dz = \int_{-R}^{R} f(x)dx + \int_{II} f(z)dz$$
$$= 2\pi i \sum_i \text{Res}\,(f(z), a_i),\qquad(5.71)$$

where a_i's are poles in the upper half-plane.

When $R \to \infty$,

$$\int_{-R}^{R} f(x)dx \to \int_{-\infty}^{\infty} f(x)dx,\qquad(5.72)$$

and

$$\left| \int_{II} f(z)dz \right| \le \left| \int_{II} \frac{P(z)}{Q(z)}dz \right|$$
$$\le \frac{1}{R^2} \left| \int_0^{\pi} Rd\theta \right|$$
$$= \mathcal{O}\left(\frac{1}{R}\right) \to 0,\qquad(5.73)$$

where we used the condition of Eq. (5.69). Therefore, it follows

$$\int_{-\infty}^{\infty} f(x)dx = 2\pi i \sum_i \text{Res}\,(f(z), a_i).\qquad(5.74)$$

$\square$

Example 1
Evaluate

$$\int_{-\infty}^{\infty} \frac{dx}{1+x^2}.\qquad(5.75)$$

The integrand satisfies the two conditions (the difference in the polynomial order is 2 and $z = \pm i$ are not on the real axis. Note that only $z = i$ is in the upper half-plane.). Therefore, we have

$$\int_{-\infty}^{\infty} \frac{dx}{1+x^2} = 2\pi i \, \mathrm{Res}\left(\frac{1}{1+z^2}; i\right)$$

$$= 2\pi i \, \frac{1}{2z}\Big|_{z=i}$$

$$= \pi. \tag{5.76}$$

Of course, this result can be also obtained by directly integrating $1/(1+x^2)$ ($=\arctan x$).

Example 2

Evaluate

$$\int_{-\infty}^{\infty} \frac{dx}{1+x^4} dx. \tag{5.77}$$

The difference in the degree of $P(x)/Q(x)$ is 4 and all the zeros to $1 + z^4 = 0$ are off the real axis. The zeros for $z^4 + 1 = 0$ are

$$z = e^{\pi i/4 + n\pi i/2}, \qquad n = 0, 1, 2, 3 \tag{5.78}$$

of which the zeros for $n = 0$ and $n = 1$ are in the upper half-plane. Therefore,

$$\int_{-\infty}^{\infty} \frac{dx}{1+x^4} = 2\pi i \sum_{\text{(Upper half plane)}} \mathrm{Res}\left(\frac{1}{z^4+1}; a_i\right)$$

$$= 2\pi i \left(\mathrm{Res}\left(\frac{1}{z^4+1}; e^{\pi i/4}\right) + \mathrm{Res}\left(\frac{1}{z^4+1}; e^{3\pi i/4}\right)\right)$$

$$= 2\pi i \left(\frac{1}{4z^3}\Big|_{z=e^{\pi i/4}} + \frac{1}{4z^3}\Big|_{z=e^{3\pi i/4}}\right)$$

$$= \frac{2\pi i}{4}\left(\frac{1}{e^{3\pi i/4}} + \frac{1}{e^{\pi i/4}}\right)$$

$$= \frac{\pi i}{2}\left(e^{-3\pi i/4} + e^{-\pi/4i}\right)$$

$$= \frac{\pi i}{2}\left(-\sqrt{2}i\right)$$

$$= \frac{\pi}{\sqrt{2}}. \tag{5.79}$$

Example 3

Evaluate

$$\int_{-\infty}^{\infty} \frac{x^2}{1+x^4} dx. \tag{5.80}$$

This integral is similar to Example 2.

$$\int_{-\infty}^{\infty} \frac{x^2}{1+x^4} dx = 2\pi i \sum_{\text{(Upper half plane)}} \text{Res}\left(\frac{z^2}{z^4+1}; a_i\right)$$

$$= 2\pi i \left(\frac{z^2}{4z^3}\Big|_{z=\exp(\pi i/4)} + \frac{z^2}{4z^3}\Big|_{z=\exp(3\pi i/4)}\right)$$

$$= \frac{\pi i}{2}\left(\frac{1}{z}\Big|_{z=\exp(\pi i/4)} + \frac{1}{z}\Big|_{z=\exp(3\pi i/4)}\right)$$

$$= \frac{\pi i}{2}\left(e^{-\pi i/4} + e^{-3\pi i/4}\right)$$

$$= \frac{\pi i}{2}\left(-\sqrt{2}i\right)$$

$$= \frac{\pi}{\sqrt{2}}. \tag{5.81}$$

Second Type Variation

As a variation of the second type of improper integrals, if $f(x)$ is multiplied by $\sin x$ or $\cos x$, the following holds:

$$\int_{-\infty}^{\infty} f(x)e^{ix} dx = 2\pi i \sum_{\text{(Upper half plane)}} \text{Res}(f(z)e^{iz}; a_i), \tag{5.82}$$

where a_i's are the poles of $f(z)e^{iz}$ taken from the upper half-plane if

1. $\deg Q(x)/P(x) \geq 1$;
2. There is no x such that $Q(x) = 0$.

The proof can be done similar to that for the second type.

Example 1

Evaluate

$$\int_{-\infty}^{\infty} \frac{\cos x}{1+x^2} dx. \tag{5.83}$$

Note that this integral is the real part of

$$\int_{-\infty}^{\infty} \frac{e^{ix}}{1+x^2} dx. \tag{5.84}$$

Therefore,

$$\int_{-\infty}^{\infty} \frac{e^{ix}}{1+x^2} dx = 2\pi i \text{ Res}\left(\frac{e^{iz}}{1+z^2}; i\right)$$

$$= 2\pi i \frac{e^{i^2}}{2i}$$

$$= \frac{\pi}{e}. \tag{5.85}$$

Example 2

Evaluate

$$I \equiv \int_0^\infty \frac{x \sin x}{x^2 + a^2} dx. \tag{5.86}$$

Note that this integral is one half of

$$\int_{-\infty}^\infty \frac{x \sin x}{a^2 + x^2} dx. \tag{5.87}$$

Therefore,

$$
\begin{aligned}
I &= \frac{1}{2} \Im \left(\int_{-\infty}^\infty \frac{x e^{ix}}{x^2 + a^2} dx \right) \\
&= \frac{1}{2} \Im \left(2\pi i \operatorname{Res} \left(\frac{z e^{iz}}{z^2 + a^2} ; ai \right) \right) \\
&= \frac{1}{2} \Im \left(2\pi i \frac{ai e^{i(ai)}}{2ai} \right) \\
&= \frac{1}{2} \Im \left(\pi i e^{-a} \right) \\
&= \frac{\pi}{2e^a}.
\end{aligned} \tag{5.88}
$$

What if $z = a$ is on the Real Axis?

One of the requirements for the integrals of the second type is that there is no x such that $Q(x) = 0$, i.e., the denominator of $f(z)$ should not have roots on the real axis.

However, if $z = a$ is a first order pole and on the real axis, it can be shown that $z = a$ acts as one half of a pole in the upper half-plane for the integration, i.e.,

$$\int_{-\infty}^\infty f(x)dx = 2\pi i \sum_{\text{(Upper half plane)}} \operatorname{Res}(f(z); z_i) + \pi i \operatorname{Res}(a). \tag{5.89}$$

Proof Assume that $z = a$ is a first order pole on the real axis and there are no other poles in the upper half-plane. Consider

$$\oint f(z)dz, \tag{5.90}$$

where the integral path is shown in Fig. 5.7 that encloses $z = a$ inside the path.

Fig. 5.7 A pole is on the real axis and the integral path

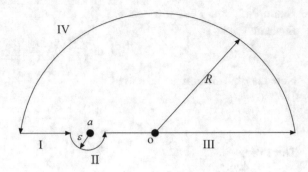

By the residue theorem, we have

$$\oint f(z)dz = 2\pi i \text{Res}(f(z); a),\tag{5.91}$$

as $z = a$ is inside the integral path. The total integral path can be split into four segments as in Fig. 5.7.

1. Paths I + III: As $R \to \infty$ and $\epsilon \to 0$, the integral along Paths I and III approaches an integral from $-\infty$ to ∞ as

$$\int_{I+III} \to \int_{-\infty}^{\infty} f(x)dx.\tag{5.92}$$

2. Path II: Along Path II which is a semi-circle, we can set $z - a = \epsilon e^{i\theta}$ where θ varies from π to 2π and $dz = \epsilon e^{i\theta} id\theta$. Recalling that $f(z)$ can be expanded in the neighborhood of $z = a$ (a first order pole) as

$$f(z) = \frac{c_{-1}}{z - a} + c_0 + c_1(z - a) + c_2(z - a)^2 + \cdots,\tag{5.93}$$

it follows

$$\int_{II} f(z)dz = \int_{II} \left(\frac{c_{-1}}{z - a} + c_0 + c_1(z - a) + c_2(z - a)^2 + \cdots \right) dz$$

$$= \int_{\pi}^{2\pi} (\frac{c_{-1}}{\epsilon e^{i\theta}} + c_0 + c_1 \epsilon e^{i\theta} + \cdots)\epsilon e^{i\theta} id\theta$$

$$= c_{-1} i \int_{\pi}^{2\pi} d\theta + c_0 \int_{\pi}^{2\pi} \epsilon e^{i\theta} id\theta + \cdots$$

$$= \pi i\, c_{-1}\tag{5.94}$$

as $\epsilon \to 0$.

Fig. 5.8 Alternative path when a pole is on the real axis and the integral path

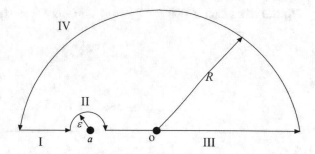

3. Path IV: As $R \to \infty$, it can be shown that the integral along Path IV approaches 0.

Thus,

$$\int_{-\infty}^{\infty} f(x)dx + \pi i c_{-1} = 2\pi i \text{Res}(f(z); a), \tag{5.95}$$

or

$$\int_{-\infty}^{\infty} f(x)dx = \pi i \text{Res}(f(z); a). \tag{5.96}$$

□

The same conclusion can be drawn by selecting an integral path shown in Fig. 5.8. The range of θ can be modified as $\theta = \pi$ to $\theta = 0$. There is no general formula available when $z = a$ is other than a first order pole.

Example 1

Evaluate

$$I \equiv \int_{-\infty}^{\infty} \frac{\sin x}{x}dx. \tag{5.97}$$

$$I = \Im \int_{-\infty}^{\infty} \frac{e^{ix}}{x}dx$$

$$= \Im \left(\pi i \times \text{Res}\left(\frac{e^{iz}}{z}; 0 \right) \right)$$

$$= \Im(\pi i \times 1)$$

$$= \pi, \tag{5.98}$$

where $\Im(f)$ is the imaginary part of f. Note that $x = 0$ is not a singular point for $\frac{\sin x}{x}$ but is the singular point for $\frac{e^{ix}}{x}$.

Fig. 5.9 The graph of $f(x)$

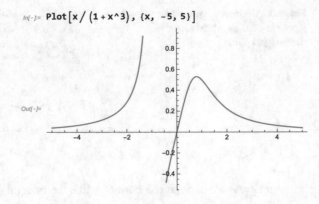

In[·]:= **Plot[x/(1+x^3), {x, -5, 5}]**

Out[·]=

Example 2
Evaluate

$$I \equiv \int_{-\infty}^{\infty} \frac{x}{1+x^3} dx. \tag{5.99}$$

This integral is convergent if we take the principal value around $x = -1$ as shown in Fig. 5.9.

The three poles are found as $z = -1$, $z = e^{\pi i/3}$ and $z = e^{5\pi i/3}$ by solving $z^3 = -1$, each of which is a first order pole. The pole, $z = -1$, is on the x-axis while $z = e^{\pi i/3}$ is in the upper half-plane. Thus, we have

$$I = 2\pi i \operatorname{Res}\left(\frac{z}{1+z^3}; e^{\pi i/3}\right) + \pi i \operatorname{Res}\left(\frac{z}{1+z^3}; -1\right)$$

$$= 2\pi i \left.\frac{z}{3z^2}\right|_{z \to \exp(\pi i/3)} + \pi i \left.\frac{z}{3z^2}\right|_{z \to -1}$$

$$= 2\pi i \frac{e^{\pi i/3}}{3e^{2\pi i/3}} + \pi i \frac{(-1)}{3(-1)^2}$$

$$= \frac{\pi i}{3}\left(2e^{-\pi i/3} - 1\right)$$

$$= \frac{\pi i}{3}\left(1 - \sqrt{3}i - 1\right)$$

$$= \frac{\sqrt{3}}{3}\pi. \tag{5.100}$$

Example 3
Evaluate

$$I \equiv \int_{-\infty}^{\infty} \frac{\sin x}{x(x^2 - a^2)} dx. \tag{5.101}$$

This integral is convergent only if the principal value is taken as shown in Fig. 5.10. Note that $z = 0$ is not singular for $f(x) = \sin x/(x(x^2 - a^2))$ although it is singular for

Fig. 5.10 The graph of $f(x)$

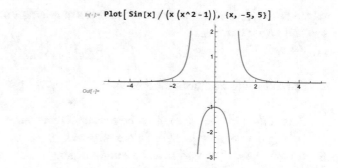

$\ln[\cdot]:=$ `Plot[Sin[x] / (x (x^2 - 1)), {x, -5, 5}]`

$Out[\cdot]:=$

$e^{iz}/(z(z^2 - a^2))$. Therefore, the residues at all the poles of $z = 0, \pm a$ are to be computed (with a half weight).

$$
\begin{aligned}
I &= \Im\left(\int_{-\infty}^{\infty} \frac{e^{ix}}{x(x^2 - a^2)}\,dx\right) \\
&= \Im\left(\pi i \times \operatorname{Res}\left(\frac{e^{iz}}{z(z^2 - a^2)}; -a\right) + \pi i \times \operatorname{Res}\left(\frac{e^{iz}}{z(z^2 - a^2)}; 0\right)\right. \\
&\quad \left. +\pi i \times \operatorname{Res}\left(\frac{e^{iz}}{z(z^2 - a^2)}; a\right)\right) \\
&= \Im\left(\pi i \times \frac{e^{iz}}{3z^2 - a^2}\Big|_{z\to -a} + \pi i \times \frac{e^{iz}}{3z^2 - a^2}\Big|_{z\to 0} + \pi i \times \frac{e^{iz}}{3z^2 - a^2}\Big|_{z\to a}\right) \\
&= \pi\Im\left(\pi i \times \frac{e^{-ia}}{3(-a)^2 - a^2} + \pi i \times \frac{1}{-a^2} + \pi i \times \frac{e^{ia}}{3a^2 - a^2}\right) \\
&= \pi\Im\left(\pi i \times \frac{e^{-ia}}{2a^2} - \pi i \times \frac{1}{a^2} + \pi i \times \frac{e^{ia}}{2a^2}\right) \\
&= \pi\Im\left(\frac{\cos a - 1}{a^2}i\right) \\
&= \frac{\cos a - 1}{a^2}\pi.
\end{aligned}
\tag{5.102}
$$

5.4.3 Third Type

The third type of integrals differs from the first and second types in that the multi-valuedness of complex functions must be taken into account. Integrals belonging to the third type vary in formats, but all of them must include functions that are multi-valued. In general, integrals of the third type contain either x^α where α is a non-integer or $\log x$ and the integral path is typically from 0 to ∞ as

$$\int_0^\infty f(x^\alpha, \log x)dx. \qquad (5.103)$$

The following conditions must be met:

1. Part of the function, $f(x)$, must be a rational function.
2. $|x^2 f(x)| < \infty$ must be held for convergence.
3. If there is an x^α term, α must be a non-integer.

A typical integral path that works for many integrals of the third type is shown in Fig. 5.11.

Example 1

Evaluate

$$I = \int_0^\infty \frac{x^\alpha}{1+x}dx, \qquad (5.104)$$

where α is a non-integer.

Instead of the integral given, consider a complex integral defined as

$$\oint \frac{z^\alpha}{1+z}dz, \qquad (5.105)$$

where the integral path is a contour shown in Fig. 5.11. The integrand, $z^\alpha/(1+z)$, has a first order pole at $z = e^{\pi i}$ (note in polar form). Although $z = 0$ is a branch point, it is outside the integral path and has no effect on the contour integral. From the residue theorem,

$$\oint_{I+II+III+IV} \frac{z^\alpha}{1+z}dz = 2\pi i \, \mathrm{Res}(f; e^{\pi i})$$

$$= 2\pi i e^{\pi i \alpha}. \qquad (5.106)$$

Fig. 5.11 A typical integral
path for the third type

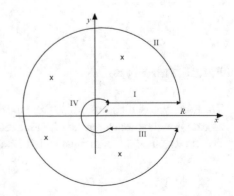

Note that we used the alternative formula to obtain the residue at the first order pole $(h(a)/g'(a))$. Note also that $z = e^{\pi i}$ was used rather than $z = -1$ because of the presence of the branch cut $((-1)^\alpha$ is ambiguous). The integral along each segment is evaluated as

1. Along Path I, $z = x$ and $dz = dx$. As $R \to \infty$ and $\epsilon \to 0$, the integral range is asymptotically expanded from (ϵ, R) to $(0, \infty)$. Therefore,

$$\int_I \to \int_0^\infty \frac{x^\alpha}{1+x} dx. \tag{5.107}$$

2. Along Path II $(z = Re^{i\theta})$, it can be shown that as $R \to \infty$, the integral approaches 0.

$$\int_{II} \to 0. \tag{5.108}$$

3. Along Path III, we can set $z = e^{2\pi i} x$ after one full rotation (2π). Note that $dz = e^{2\pi i} dx$ and $1 + z = 1 + e^{2\pi i} x = 1 + x$. Only $z^\alpha = (e^{2\pi i} x)^\alpha = e^{2\pi i \alpha} x^\alpha$ is affected by the factor, $e^{2\pi i}$.

$$\int_{III} \to \int_\infty^0 \frac{(xe^{2\pi i})^\alpha}{1 + xe^{2\pi i}} dx$$
$$= -e^{2\pi i \alpha} \int_0^\infty \frac{x^\alpha}{1+x} dx. \tag{5.109}$$

4. Along Path IV $(z = \epsilon e^{i\theta})$, it can be shown that this integral approaches 0 as $\epsilon \to 0$.

$$\int_{IV} \to 0. \tag{5.110}$$

Therefore,

$$\int_{I+II+III+IV} \to I + 0 + (-e^{2\pi i \alpha} I) + 0$$
$$= (1 - e^{2\pi i \alpha}) I. \tag{5.111}$$

Equating Eq. (5.106) with Eq. (5.111), we obtain

$$(1 - e^{2\pi i \alpha}) I = 2\pi i e^{\pi i \alpha}, \tag{5.112}$$

from which

$$I = \frac{2\pi i e^{\pi i \alpha}}{1 - e^{2\pi i \alpha}}$$

$$= \frac{2\pi i}{e^{-\pi i \alpha} - e^{\pi i \alpha}}$$

$$= \frac{2\pi i}{-2i \sin \pi \alpha}$$

$$= -\frac{\pi}{\sin \pi \alpha}. \tag{5.113}$$

Example 2

Evaluate

$$\int_0^\infty \frac{\text{Log}\, x}{1 + x^2} dx. \tag{5.114}$$

It appears odd at first but consider the following integral[5]:

$$\oint \frac{(\log z)^2}{1 + z^2} dz, \tag{5.115}$$

where the integral path is the same as in Fig. 5.11. First, from the residue theorem,

$$\oint \frac{(\log z)^2}{1 + z^2} dz = 2\pi i \left(\text{Res}(f(z); e^{\pi i/2}) + \text{Res}(f(z); e^{3\pi i/2}) \right)$$

$$= 2\pi i \left(\frac{(\pi i/2)^2}{2i} + \frac{(3\pi i/2)^2}{(-2i)} \right)$$

$$= 2\pi^3. \tag{5.116}$$

Note that we used $e^{\pi i/2}$ and $e^{3\pi i/2}$ rather than i and $-i$ for the representation of each singular point.

On the other hand, the contour above can be split into four segments as in Example 1. The integral along each segment is evaluated as

1. Along Path I, $z = x$ and $dz = dx$. As $R \to \infty$ and $\epsilon \to 0$, the integral range is asymptotically expanded from (ϵ, R) to $(0, \infty)$. Therefore,

$$\int_I \to \int_0^\infty \frac{(\text{Log}\, x)^2}{1 + x^2} dx. \tag{5.117}$$

2. Along Path II ($z = Re^{i\theta}$), it can be shown that as $R \to \infty$, the integral approaches 0.

$$\int_{II} \to 0. \tag{5.118}$$

---

[5] We use $\log z$ when z is a complex number and $\text{Log}\, x$ when x is a real positive number.

3. Along Path III, we can set $z = e^{2\pi i} x$ after one full rotation (2π). Note that $dz = e^{2\pi i} dx$ and $1 + z^2 = 1 + e^{4\pi i} x^2 = 1 + x^2$. Only $\log z = \log(e^{2\pi i} x) = \text{Log}\, x + 2\pi i$ is affected by the factor, $e^{2\pi i}$:

$$
\int_{III} \to \int_{\infty}^{0} \frac{(\log(xe^{2\pi i}))^2}{1 + x^2} dx
$$

$$
= \int_{\infty}^{0} \frac{(\text{Log}\, x)^2 + 4\pi i \text{Log}\, x - 4\pi^2}{1 + x^2} dx
$$

$$
= -\int_{0}^{\infty} \frac{(\text{Log}\, x)^2}{1 + x^2} dx - 4\pi i \int_{0}^{\infty} \frac{\text{Log}\, x}{1 + x^2} dx + 4\pi^2 \int_{0}^{\infty} \frac{1}{1 + x^2} dx.
$$

(5.119)

4. Along Path IV $(z = \epsilon e^{i\theta})$, it can be shown that this integral approaches 0 as $\epsilon \to 0$:

$$
\int_{IV} \to 0.
$$

(5.120)

Thus, equating both sides, we have

$$
2\pi^3 = \int_{0}^{\infty} \frac{(\text{Log}\, x)^2}{1 + x^2} dx + 0
$$

$$
+ \left(-\int_{0}^{\infty} \frac{(\text{Log}\, x)^2}{1 + x^2} dx - 4\pi i \int_{0}^{\infty} \frac{\text{Log}\, x}{1 + x^2} dx + 4\pi^2 \int_{0}^{\infty} \frac{1}{1 + x^2} dx \right) + 0
$$

$$
= -4\pi i \int_{0}^{\infty} \frac{\text{Log}\, x}{1 + x^2} dx + 4\pi^2 \int_{0}^{\infty} \frac{1}{1 + x^2} dx
$$

$$
= -4\pi i \int_{0}^{\infty} \frac{\text{Log}\, x}{1 + x^2} dx + 4\pi^2 \frac{\pi}{2}
$$

$$
= -4\pi i \int_{0}^{\infty} \frac{\text{Log}\, x}{1 + x^2} dx + 2\pi^3.
$$

(5.121)

Therefore,

$$
\int_{0}^{\infty} \frac{\text{Log}\, x}{1 + x^2} dx = 0,
$$

(5.122)

not because of the Cauchy theorem.

5.4.4 *Mathematica* **Code**

Although the residue theorem is important in evaluating certain types of integrals, its emphasis is probably less important today as computer algebra systems including *Mathematica* can evaluate many integrations automatically. Below is *Mathematica* code for the integrals in the preceding examples.

In[ ]:= `Integrate[1/ (3 + Sin[th]), {th, 0, 2 Pi}]`

Out[ ]= $\dfrac{\pi}{\sqrt{2}}$

In[ ]:= `Integrate[Cos[th]^6, {th, 0, 2 Pi}]`

Out[ ]= $\dfrac{5\pi}{8}$

In[ ]:= `Integrate[1/ (1 + x^2), {x, -Infinity, Infinity}]`

Out[ ]= π

In[ ]:= `Integrate[1/ (1 + x^4), {x, -Infinity, Infinity}]`

Out[ ]= $\dfrac{\pi}{\sqrt{2}}$

In[ ]:= `Integrate[x^2/ (1 + x^4), {x, -Infinity, Infinity}]`

Out[ ]= $\dfrac{\pi}{\sqrt{2}}$

In[ ]:= `Integrate[Cos[x]/ (1 + x^2), {x, -Infinity, Infinity}]`

Out[ ]= $\dfrac{\pi}{e}$

In[ ]:= `Integrate[x Sin[x]/ (x^2 + a^2), {x, -Infinity, Infinity}]`

Out[ ]= $\boxed{e^{-a}\pi \ \ \text{if} \ \ \text{Re}[a] > 0}$

In[ ]:= `Integrate[Sin[x]/x, {x, -Infinity, Infinity}]`

Out[ ]= π

In[ ]:= `Integrate[x/ (1 + x^3), {x, -Infinity, Infinity}, PrincipalValue → True]`

Out[ ]= $\dfrac{\pi}{\sqrt{3}}$

In[ ]:= `Integrate[Sin[x]/x/ (x^2 - a^2), {x, -Infinity, Infinity}, PrincipalValue → True]`

Out[ ]= $\boxed{\dfrac{\pi\,(-1 + \text{Cos}[a])}{a^2} \ \ \text{if} \ \ a \in \mathbb{R}}$

In[ ]:= `Integrate[x^α/ (1 + x^2), {x, 0, Infinity}]`

Out[ ]= $\boxed{\dfrac{1}{2}\pi\,\text{Sec}\!\left[\dfrac{\pi\alpha}{2}\right] \ \ \text{if} \ \ -1 < \text{Re}[\alpha] < 1}$

In[ ]:= `Integrate[Log[x]/ (1 + x^2), {x, 0, Infinity}]`

Out[ ]= 0

For the integrals of $\frac{x}{1+x^3}$ and $\frac{\sin x}{x(x^2-a^2)}$, an option `PrincipalValue->True` has to be added. Otherwise, the integrals are divergent.

5.5   Problems

1. Evaluate

$$\int_0^{2\pi} \frac{1}{a + \cos\theta} d\theta, \quad a > 0. \tag{5.123}$$

2. Evaluate

$$\int_{-\infty}^{\infty} \frac{x^2}{x^6 + 1} dx. \tag{5.124}$$

3. Evaluate the following integral:

$$\int_0^{\infty} \frac{dx}{x^2 + 3x + 2}, \tag{5.125}$$

by considering

$$\oint \frac{\log z}{z^2 + 3z + 2} dz, \tag{5.126}$$

using the same integral path as in Fig. 5.11.

4. Evaluate

$$\oint \frac{1}{z - 1} dz, \tag{5.127}$$

where the integral path is a rectangle whose corners are $-1, +2, 2 + i$ and $-1 + i$.

5. Compute the residue of

$$\frac{e^z - 1}{z^5}, \tag{5.128}$$

at $z = 0$.

6. Evaluate

$$\int_{-\infty}^{\infty} \frac{x \sin x}{(x^2 + 1)(x^2 + 4)} dx. \tag{5.129}$$

Applications to Engineering Problems

6

In this chapter, applications of complex variable theory to engineering are presented. Many differential equations that arise in engineering are variations of the Laplace/Poisson type of equations. As both the real and imaginary parts of an analytic function automatically satisfy the Laplace equations, it is natural that an analytic function finds its way into a solution technique for these equations.

We consider boundary value problems in heat conduction, fluid mechanics and solid mechanics, but equations in other disciplines can be also handled in a similar manner. We first introduce conformal mapping as a way to change the original boundary shape to a simpler boundary while keeping the format of the Laplace equation. We then consider general methods for solving 2-D Laplace and bi-harmonic equations using Taylor/Laurent series for analytic functions, Many problems in steady-state heat conduction and 2-D elasticity can be solved by this method. Complex variables also play an important role in fluid mechanics. An analytic function is almost synonymous to a potential flow, and we show how various potential flows can be realized by a combination of simple analytic functions.

Mathematica is used throughout to facilitate computations without which manual calculation is very difficult. It also helps to visualize streamlines of potential flows.

6.1 Conformal Mapping

A conformal mapping is defined as a mapping from $z = x + yi$ to another complex number $w = u + vi$ through

$$w = f(z), \tag{6.1}$$

© The Author(s), under exclusive license to Springer Nature Switzerland AG 2022
S. Nomura, *Complex Variables for Engineers with Mathematica*,
Synthesis Lectures on Mechanical Engineering,
https://doi.org/10.1007/978-3-031-13067-0_6

Fig. 6.1 Conformal mapping

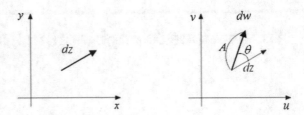

where $f(z)$ is an analytic function of z.[1] By taking the total derivative of Eq. (6.1) at $z = z_0$, we have

$$dw = f'(z_0)dz = Ae^{i\theta}dz, \tag{6.2}$$

where A and θ are the amplitude and the argument of $f'(z_0)$, respectively. Equation (6.2) implies that the line element, dz, in the z-plane is stretched by A and rotated by θ in the w-plane as shown in Fig. 6.1. When $f'(z_0) = 0$, dz is mapped to a point and conformal mapping fails.

It is clear that when two curves that met at $z = z_0$ in the z-plane are mapped to the w-plane via a conformal mapping, the angle between the curves is preserved, hence, the name, *conformal mapping*.

A coordinate transformation from (x, y) to (u, v) offers an opportunity that an irregular boundary in the (x, y) plane could be mapped to a regular boundary in the (u, v) plane in which one can solve the equation with ease. There is, of course, no guarantee that the Laplace equation in (x, y) remains the Laplace equation in (u, v). However, it can be shown that if the Cauchy-Riemann equations $(u_x = v_y, u_y = -v_x)$ are held between (x, y) and (u, v), the Laplace equation in (x, y) remains the Laplace equation in (u, v).

Proof Consider a coordinate transformation from (x, y) to (u, v). We want to express the Laplacian ($\Delta \equiv \frac{\partial^2}{\partial x^2} + \frac{\partial^2}{\partial y^2}$) in terms of (u, v). Using the chain differentiation rule combined with the Cauchy-Riemann equations, we have

$$\begin{aligned} \frac{\partial}{\partial x} &= \frac{\partial u}{\partial x}\frac{\partial}{\partial u} + \frac{\partial v}{\partial x}\frac{\partial}{\partial v} \\ &= u_x \partial_u - u_y \partial_v, \end{aligned} \tag{6.3}$$

where u_x and ∂u are shorthands for $\frac{\partial u}{\partial x}$ and $\frac{\partial}{\partial u}$, respectively, and one of the Cauchy-Riemann equations, $u_y = -v_x$, was used. Differentiating Eq. (6.3) with respect to x yields

$$\frac{\partial^2}{\partial x^2} = u_{xx}\partial_u + u_x\partial_x\partial_u - u_{yx}\partial_v - u_y\partial_x\partial_v. \tag{6.4}$$

Repeating the above for y, we have

---

[1] Conformal mapping and analytic functions are about the same.

$$\frac{\partial}{\partial y} = \frac{\partial u}{\partial y}\frac{\partial}{\partial u} + \frac{\partial v}{\partial y}\frac{\partial}{\partial v}$$
$$= u_y \partial_u + u_x \partial_v,$$
(6.5)

and

$$\frac{\partial^2}{\partial y^2} = u_{yy}\partial_u + u_y\partial_y\partial_u + u_{xy}\partial_v + u_x\partial_y\partial_v.$$
(6.6)

Using

$$\partial_x \partial_u = (u_x \partial_u - u_y \partial_v)\partial_u = u_x \partial_{uu} - u_y \partial_{uv},$$
(6.7)
$$\partial_x \partial_v = (u_x \partial_u - u_y \partial_v)\partial_v = u_x \partial_{uv} - u_y \partial_{vv},$$
(6.8)
$$\partial_y \partial_u = (u_y \partial_u + u_x \partial_v)\partial_u = u_y \partial_{uu} + u_x \partial_{vu},$$
(6.9)
$$\partial_y \partial_v = (u_y \partial_u + u_x \partial_v)\partial_v = u_y \partial_{uv} + u_x \partial_{vv},$$
(6.10)

it follows

$$\frac{\partial^2}{\partial x^2} + \frac{\partial^2}{\partial y^2} = (u_x^2 + u_y^2)\left(\frac{\partial^2}{\partial u^2} + \frac{\partial^2}{\partial v^2}\right)$$
$$= |f'(z)|^2 \left(\frac{\partial^2}{\partial u^2} + \frac{\partial^2}{\partial v^2}\right),$$
(6.11)

where we used

$$\frac{\partial f}{\partial x} = \frac{\partial z}{\partial x}\frac{\partial f}{\partial z} + \frac{\partial \bar{z}}{\partial x}\frac{\partial f}{\partial \bar{z}} = f'(z),$$
(6.12)

and

$$|f'(z)|^2 = \left|\frac{\partial f}{\partial x}\right|^2 = |u_x + iv_x|^2 = u_x^2 + v_x^2 = u_x^2 + u_y^2.$$
(6.13)

Therefore, if $f'(z) \neq 0$, the Laplace equation in (x, y) is also the Laplace equation in (u, v). □

Conformal mapping can be visualized by the following *Mathematica* code as shown in Figs. 6.2 and 6.3. In Fig. 6.2, a conformal mapping, $w = z^2$, is shown in which the loci of $u = const$ and $v = const$ are plotted in the z-plane as a set of mutually orthogonal curves.

By changing z^2 in the code, any conformal mapping can be visualized. A graph for $w = 1/z$ is shown in Fig. 6.3.

6.1.1 Solving Laplace Equation by Conformal Mapping

The Laplace equation remains the Laplace equation through conformal mapping. Therefore, it is possible that an irregular region in the z-plane could be mapped to a regular region in the

Fig. 6.2 Visualization of z^2

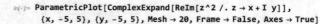

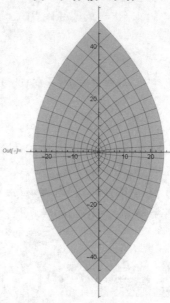

Fig. 6.3 Visualization of $\frac{1}{z}$

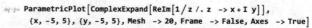

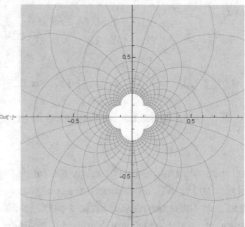

w-plane through conformal mapping and the Laplace equation can be solved with ease in the w-plane. We show two examples where conformal mapping helps to solve steady-state heat conduction problems. Detailed discussion on conformal mapping as to what type of mapping should be chosen is found in [10].

Example 1

In this example, consider a steady-state heat conduction in a half-plane outside a cylinder with a radius of $1/2$ shown in Fig. 6.4. The temperature distribution, $T(x, y)$, is subject to the Laplace equation as

$$\Delta T = 0. \tag{6.14}$$

The boundary condition is that $T = 100\,^\circ\text{C}$ along $x = 0$ and $T = 50\,^\circ\text{C}$ along the wall of the cylinder whose equation in the (x, y) plane is expressed as

$$\left| z - \frac{1}{2} \right| = \frac{1}{2}. \tag{6.15}$$

Among all possible conformal mappings from the z-plane to w-plane, we choose

$$w = \frac{1}{z}. \tag{6.16}$$

To transform the shaded region into the w-plane, the boundary lines that define the shaded region can be mapped to the w-plane.

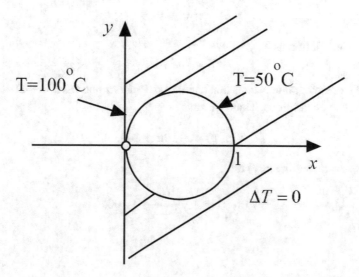

Fig. 6.4 Temperature distribution around a cylinder in a half-plane

1. $x = 0$ (the y-axis)

   The condition, $x = 0$, is equivalent to $z + \bar{z} = 0$ which can be rewritten in terms of w as

$$
\begin{aligned}
z + \bar{z} &= \frac{1}{w} + \frac{1}{\bar{w}} \\
&= \frac{w + \bar{w}}{w\bar{w}} \\
&= 0.
\end{aligned}
\tag{6.17}
$$

   Therefore, $z + \bar{z} = 0$ is mapped to $w + \bar{w} = 0$, i.e., the y-axis is mapped to the v-axis.

2. $\left| z - \frac{1}{2} \right| = \frac{1}{2}$ (a circle centered at $z = \frac{1}{2}$ with a radius of $\frac{1}{2}$)

   Substituting $z = 1/w$ yields

$$
\left| \frac{1}{w} - \frac{1}{2} \right| = \frac{1}{2},
\tag{6.18}
$$

   or equivalently,

$$
|w - 2| = |w - 0|,
\tag{6.19}
$$

   which represents the midline between $w = 2$ and $w = 0$, i.e., $u = 1$.

Thus, the domain in the x-y plane is mapped to an infinite strip surrounded by $u = 0$ and $u = 1$ in the u-v plane as shown in Fig. 6.5. As this region is extended to $v \to \pm\infty$, T must be independent of v. Therefore, the Laplace equation in (u, v) is reduced to

$$
\frac{d^2 T}{du^2} = 0,
\tag{6.20}
$$

which yields the solution as

$$
T = au + b,
\tag{6.21}
$$

where a and b are integral constants that can be easily determined as $a = -50$ and $b = 100$ from the boundary conditions. Thus, we have

$$
T(u, v) = -50u + 100,
\tag{6.22}
$$

which can be expressed in (x, y) as[2]

$$
T(x, y) = -\frac{50x}{x^2 + y^2} + 100.
\tag{6.23}
$$

---

[2] The relation, $w = 1/z$, is expressed as $u + vi = \frac{1}{x+yi}$, or $u = \frac{x}{x^2+y^2}$, $v = -\frac{y}{x^2+y^2}$.

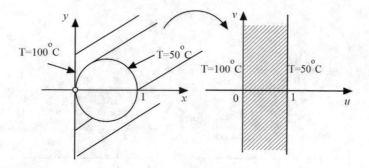

Fig. 6.5 Conformal mapping from z-plane to w-plane

Fig. 6.6 Temperature
distribution between two
parabolas

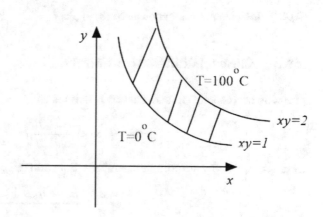

Example 2

In this example, a medium surrounded by the two curves, $xy = 1$ and $xy = 2$, shown in
Fig. 6.6 is considered. As the Dirichlet boundary condition, the temperature is prescribed
as 100°C along $xy = 2$ and 0°C along $xy = 1$. With the transformation[3] of $w = z^2$ where
$v = 2xy$, $xy = 2$ is mapped to $v = 4$ and $xy = 1$ is mapped to $v = 2$ as shown in Fig. 6.7.
Therefore, the solution for T in this region is

$$T(u, v) = 50v - 100, \tag{6.24}$$

which is fed into the x-y plane as

$$T(x, y) = 100xy - 100. \tag{6.25}$$

---

[3] $w = z^2$ was chosen as the term, xy, is half the imaginary part of $z^2 = (x + yi)^2$.

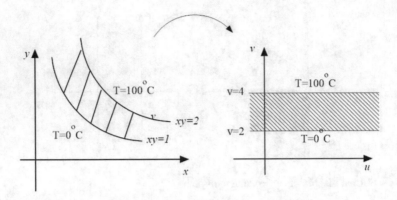

Fig. 6.7 Temperature distribution in the (u, v) plane

6.1.2   Bilinear (Möbius) Transformation

The bilinear (Möbius) transformation is defined as

$$w = \frac{az + b}{cz + d},$$

(6.26)

where a, b, c and d are complex numbers. Equation (6.26) can be rewritten as

$$w = \frac{a}{c} + \frac{bc - ad}{c^2} \frac{1}{z + \frac{d}{c}}.$$

(6.27)

It is seen from Eq. (6.27) that the bilinear transformation of Eq. (6.26) is a combination of translation ($\frac{a}{c}$ and $\frac{d}{c}$), dilatation ($\frac{bc-ad}{c^2}$) and inversion ($\frac{1}{z}$), all of which are circle preservation mappings, i.e., a circle in the z-plane (including a straight line, which is a special case of a circle by $r \to \infty$) is mapped to another circle (including a straight line) in the w-plane. This way, if the domain of the Laplace equation is defined by a combination of circles (including straight lines), such a domain can be mapped to either concentric circles or a long strip in which the Laplace equation is easier to be solved. For a complete reference of conformal mapping, see [10].

6.2   General Solution to Laplace Equation ($\Delta\phi(x, y) = 0$)

In Sect. 6.1, conformal mapping was shown to be useful for solving 2-D Laplace equations by mapping an original domain to a domain with simpler geometry. In this section, an alternative approach to solving 2-D Laplace equations (harmonic equations) is introduced using a series of analytic functions. This approach can be further extended to a solution method for 2-D bi-harmonic equations whose immediate application is found in 2-D elasticity in Sect. 6.5.

The relationship between (x, y) and $(z, \bar{z})$,

$$z = x + yi, \quad \bar{z} = x - yi \tag{6.28}$$

is equivalent to

$$x = \frac{1}{2}(z + \bar{z}), \quad y = \frac{1}{2i}(z - \bar{z}). \tag{6.29}$$

Using the chain differentiation rule,

$$\frac{\partial}{\partial z} = \frac{\partial x}{\partial z}\frac{\partial}{\partial x} + \frac{\partial y}{\partial z}\frac{\partial}{\partial y} = \frac{1}{2}\frac{\partial}{\partial x} + \frac{1}{2i}\frac{\partial}{\partial y}, \tag{6.30}$$

$$\frac{\partial}{\partial \bar{z}} = \frac{\partial x}{\partial \bar{z}}\frac{\partial}{\partial x} + \frac{\partial y}{\partial \bar{z}}\frac{\partial}{\partial y} = \frac{1}{2}\frac{\partial}{\partial x} - \frac{1}{2i}\frac{\partial}{\partial y}, \tag{6.31}$$

we have

$$\begin{aligned}
\frac{\partial}{\partial z}\frac{\partial}{\partial \bar{z}} &= \frac{1}{4}\left(\frac{\partial}{\partial x} + \frac{1}{i}\frac{\partial}{\partial y}\right)\left(\frac{\partial}{\partial x} - \frac{1}{i}\frac{\partial}{\partial y}\right) \\
&= \frac{1}{4}\left(\frac{\partial^2}{\partial x^2} + \frac{\partial^2}{\partial y^2}\right) \\
&= \frac{1}{4}\Delta.
\end{aligned} \tag{6.32}$$

Therefore, the Laplacian, Δ, is expressed by successive differentiations with respect to $\bar{z}$ and z.

A function, $\phi(x, y)$, that satisfies the Laplace equation, $\Delta\phi(x, y) = 0$, is sought. Through Eq. (6.29), $\phi(x, y)$ can be viewed as a function of z and $\bar{z}$, i.e., $\phi(z, \bar{z})$. Thus, $\Delta\phi(x, y) = 0$ is written as

$$\frac{\partial^2\phi(z, \bar{z})}{\partial z\partial \bar{z}} = 0, \tag{6.33}$$

or

$$\frac{\partial}{\partial z}\left(\frac{\partial\phi(z, \bar{z})}{\partial \bar{z}}\right) = 0. \tag{6.34}$$

Equation (6.34) is automatically satisfied if

$$\frac{\partial\phi(z, \bar{z})}{\partial \bar{z}} = f(\bar{z}), \tag{6.35}$$

where $f(\bar{z})$ is an arbitrary function of $\bar{z}$ alone. By integrating Eq. (6.35) with respect to $\bar{z}$, we have

$$\begin{aligned}
\phi(z, \bar{z}) &= \int f(\bar{z})d\bar{z} + g(z) \\
&= F(\bar{z}) + g(z),
\end{aligned} \tag{6.36}$$

where $F(\bar{z})$ is an arbitrary function of $\bar{z}$ alone and $g(z)$ is an arbitrary function of z alone. As $\phi(z, \bar{z})$ must be real, we have

$$\phi(x, y) = \Re\left(F(\bar{z}) + g(z)\right), \tag{6.37}$$

where $\Re(f(z))$ is the real part of $f(z)$.

Question: This conclusion seems to contradict the previous statement that the real and imaginary parts of any analytic function are harmonic because obviously $F(\bar{z})$ is NOT an analytic function?

Answer: The statement that f is an analytic function is really a sufficient condition for $\Delta f = 0$, not a necessary condition. A harmonic function can be analytic or non-analytic.

As a matter of fact, Eq. (6.37) is redundant as the real parts of $f(z)$ and $f(\bar{z})$ are the same. Hence, it suffices to write

$$\phi(x, y) = \Re(g(z)), \tag{6.38}$$

where $g(z)$ is an arbitrary analytic function without loss of generality. This result will be utilized in solving steady-state heat conduction problems in Sect. 6.4.

6.3   General Solution to Bi-harmonic Equation ($\Delta\Delta\phi(x, y) = 0$)

In this section, following a general solution method for the Laplace equation in Sect. 6.2, a similar approach for the bi-harmonic equation is presented. The major application for the solution of the bi-harmonic equations is in solid mechanics (2-D elasticity) using the Airy stress function in Sect. 6.5.

A function, $\phi(x, y)$, which satisfies the following equation:

$$\Delta\Delta\phi(x, y) = 0, \tag{6.39}$$

where

$$\begin{aligned}
\Delta\Delta &= \left(\frac{\partial^2}{\partial x^2} + \frac{\partial^2}{\partial y^2}\right)\left(\frac{\partial^2}{\partial x^2} + \frac{\partial^2}{\partial y^2}\right) \\
&= \frac{\partial^4}{\partial x^4} + 2\frac{\partial^4}{\partial x^2 \partial y^2} + \frac{\partial^4}{\partial y^4},
\end{aligned} \tag{6.40}$$

is called a bi-harmonic function. Solutions to Eq. (6.39) can be obtained by first solving

$$\Delta\Delta\Phi(z, \bar{z}) = 0, \tag{6.41}$$

for a complex function, $\Phi(z, \bar{z})$, and thereafter by taking the real part as

$$\phi(x, y) = \Re\big(\Phi(z, \bar{z})\big). \tag{6.42}$$

Equation (6.41) implies that $\Delta\Phi(z, \bar{z})$ is a harmonic function, i.e.,

$$\Delta\Phi(z, \bar{z}) = f_1(z) + g_1(\bar{z}), \tag{6.43}$$

where $f_1(z)$ is an arbitrary function of z alone and $g_1(\bar{z})$ is an arbitrary function of $\bar{z}$ alone. Equation (6.43) can be written as

$$\frac{\partial}{\partial z}\frac{\partial}{\partial \bar{z}}\Phi(z, \bar{z}) = \frac{1}{4}\big(f_1(z) + g_1(\bar{z})\big)$$
$$= f_2(z) + g_2(\bar{z}). \tag{6.44}$$

By integrating Eq. (6.44) with respect to z, we have

$$\frac{\partial}{\partial \bar{z}}\Phi(z, \bar{z}) = \int f_2(z)dz + g_2(\bar{z})\int 1dz$$
$$= f_3(z) + zg_2(\bar{z}) + g_4(\bar{z}). \tag{6.45}$$

By integrating again Eq. (6.45) with respect to $\bar{z}$, we have

$$\Phi(z, \bar{z}) = f_3(z)\int 1d\bar{z} + z\int g_2(\bar{z})d\bar{z} + \int g_4(\bar{z})d\bar{z}$$
$$= \bar{z}f_3(z) + zg_3(\bar{z}) + g_5(\bar{z}) + f_4(z). \tag{6.46}$$

Therefore, $\phi(x, y)$ can be expressed as

$$\phi(x, y) = \Re\big(\bar{z}f_1(z) + g_1(z) + zf_2(\bar{z}) + g_2(\bar{z})\big), \tag{6.47}$$

where after rearranging, f_1 and g_1 are arbitrary functions of z alone and f_2 and g_2 are arbitrary functions of $\bar{z}$ alone. In reality, Eq. (6.47) is redundant as the real parts of $\bar{z}f(z) + g(z)$ and $zf(\bar{z}) + g(\bar{z})$ are the same. Hence, without any loss of generality, we can write a solution, $\phi(x, y)$, to the bi-harmonic equation as

$$\phi(x, y) = \Re\big(\bar{z}f(z) + g(z)\big). \tag{6.48}$$

Example

As an example of generating a function that satisfies the bi-harmonic equation, choose $f(z)$ and $g(z)$ rather arbitrarily as

$$f(z) = \frac{1}{1 + z}, \quad g(z) = \frac{1}{z^2}. \tag{6.49}$$

A bi-harmonic function, $\phi(x, y)$, can be automatically constructed as

$$\phi(x, y) = \Re(\bar{z} f(z) + g(z))$$
$$= \frac{2x^2}{\left(x^2 + y^2\right)^2} - \frac{1}{x^2 + y^2} + \frac{(x + 1)(2x + 1)}{(x + 1)^2 + y^2} - 1. \tag{6.50}$$

Mathematica can automate the process above with the code listed as follows:

```
In[ ]:= f = 1 / (1 + z); g = 1 / z^2; z = x + I y; zbar = x - I y;
```

```
In[ ]:= zbar f + g /. z → x + I y
```

$$Out[ ]= \frac{1}{\left(x + i\,y\right)^2} + \frac{x - i\,y}{1 + x + i\,y}$$

```
In[ ]:= phi = ComplexExpand[Re[zbar f + g /. z → x + I y]] // FullSimplify
```

$$Out[ ]= -1 + \frac{2\,x^2}{\left(x^2 + y^2\right)^2} - \frac{1}{x^2 + y^2} + \frac{1 + x}{\left(1 + x\right)^2 + y^2} \frac{1 + 2\,x}{}$$

```
In[ ]:= laplacian[f_] := D[f, {x, 2}] + D[f, {y, 2}]
```

```
In[ ]:= laplacian[laplacian[phi]] // Simplify
```

```
Out[ ]= 0
```

6.4   Heat Conduction

In heat transfer, the steady-state heat conduction without a heat source in a homogeneous[4] and isotropic[5] body is governed by the Laplace equation.

In Sect. 6.1, we showed how conformal mapping can be used to solve certain types of Laplace equations. It was necessary to choose an appropriate conformal mapping that matches the given geometry that may require experience and know-how skills.

In this section, we show that the Taylor/Laurent series of an analytic function can be also used for the solution method of the Laplace equation. This approach is often more straightforward than the conformal mapping method as one can start with expanding an unknown analytic function by the Taylor/Laurent series about a point of interest (often a singular point) and determine the unknown coefficients so as to satisfy the boundary conditions. More terms can be added as necessary until the boundary conditions are satisfied.

First, we revisit the two previous examples in Sect. 6.1.1 in which conformal mapping was used and show that the series expansion method also works for the same problems as well. We then show additional three examples using the series expansion method.

---

[4] A body whose properties remain unchanged under translation.
[5] A body whose properties are independent of the directions.

Example 1

As an example of using the Taylor/Laurent series for solving the Laplace equation, we revisit Example 1 in Sect. 6.1.1 which was solved by the conformal mapping method but this time using the series expansion method instead. We are seeking $f(z)$, an analytic function, whose real part satisfies the given boundary condition. As seen in Fig. 6.8, $z = 0$ is the singular point at which there is a discontinuity of the temperature. Therefore, it stands to reason that $f(z)$ can be expanded by the Laurent series about $z = 0$. The simplest such form is expressed as

$$f(z) = c_0 + \frac{c_1}{z}, \tag{6.51}$$

where c_0 and c_1 are unknown complex coefficients. More terms can be added if c_0 and c_1 alone cannot satisfy the prescribed boundary condition. By setting $c_0 = a + bi$ and $c_1 = c + di$, $f(z)$ is expressed as

$$f(z) = c_0 + \frac{c_1}{z}$$
$$= (a + bi) + \frac{c + di}{x + yi}. \tag{6.52}$$

The real part of $f(z)$ is

$$T = \Re(f) = a + \frac{cx + dy}{x^2 + y^2}. \tag{6.53}$$

Along $x = 0$, $T = 100$, which is translated to $a + d/y = 100$. For this to be independent of y, $a = 100$, $d = 0$. Along $x^2 + y^2 - x = 0$, $T = 50$, which is translated into $50 = 100 + c$ from which $c = -50$. Thus, we have

Fig. 6.8 Same problem as in Example 1

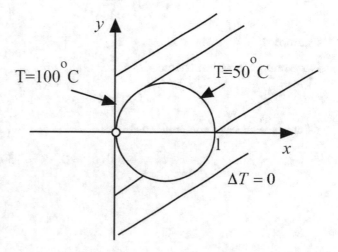

$$T = 100 - \frac{50x}{x^2 + y^2}, \tag{6.54}$$

which of course matches the solution, Eq. (6.23), by the conformal mapping method.

The following *Mathematica* code implements this procedure:

```
In[ ]:= f = (a + b I) + (c + d I) / (x + I y)
```

$$Out[\ ]= a + i\,b + \frac{c + i\,d}{x + i\,y}$$

```
In[ ]:= t1 = ComplexExpand[Re[f]]
```

$$Out[\ ]= a + \frac{c\,x}{x^2 + y^2} + \frac{d\,y}{x^2 + y^2}$$

```
In[ ]:= t1 /. x → 0
```

$$Out[\ ]= a + \frac{d}{y}$$

```
In[ ]:= sol1 = {a → 100, d → 0}
```

$$Out[\ ]= \{a → 100, d → 0\}$$

```
In[ ]:= t1 /. x^2 + y^2 → x
```

$$Out[\ ]= a + c + \frac{d\,y}{x}$$

```
In[ ]:= sol2 = {d → 0, c → -50}
```

$$Out[\ ]= \{d → 0, c → -50\}$$

```
In[ ]:= t1 /. sol1 /. sol2
```

$$Out[\ ]= 100 - \frac{50\,x}{x^2 + y^2}$$

Example 2

For Example 2, we can choose

$$f(z) = c_0 + c_1 z + c_2 z^2, \tag{6.55}$$

where $c_0 \sim c_2$ can be determined as

$$c_0 = -100, \ c_1 = 0, \ c_2 = -50i. \tag{6.56}$$

Example 3

This problem was taken from [8, (p. 1161)] solved by the conformal mapping method. We show that the series expansion method employed for Examples 1 and 2 can be also used for this problem.

The steady-state temperature distribution, $\Psi(x, y)$, is sought over the region shown in Fig. 6.9. The region is surrounded by a partial circle centered at $(x, y) = (0, 1)$ with a radius of $\sqrt{2}$ and a segment on the x-axis over $-1 < x < 1$. The boundary condition is such that $\Psi(x, y) = 0$ over the segment on the x-axis and 1 along the circle. As both $z = 1$ and $z = -1$ are singular points where Ψ is discontinuous, we assume that the analytic function, $f(z)$, whose real part represents $\Psi(x, y)$ is in the form of

$$f(z) = (c_0 + c_1 i) + (c_2 + c_3 i) \log(z - 1) + (c_4 + c_5 i) \log(z + 1), \quad (6.57)$$

where $c_0 \sim c_5$ are unknown real constants. The terms, $\log(z - 1)$ and $\log(z + 1)$, are introduced so that they can represent the multi-valuedness at $z = 1$ and $z = -1$. With $\log(z - 1)$ and $\log(z + 1)$, the derivative of $f(z)$ has terms, $1/(z - 1)$ and $1/(z + 1)$, that make the series complete. The real part of $f(z)$ is expressed from Eq. (6.57) as

$$\Psi(x, y) = c_0 - c_3 \arg(z - 1) - c_5 \arg(z + 1)$$
$$+ c_2 \log\sqrt{(x - 1)^2 + y^2} + c_4 \log\sqrt{(x + 1)^2 + y^2}. \quad (6.58)$$

On the real axis, Eq. (6.58) takes

$$\Psi(x, 0) = c_0 - c_3 \arg(x - 1) - c_5 \arg(x + 1) + c_2 \log\sqrt{(x - 1)^2} + c_4 \log\sqrt{(x + 1)^2}. \quad (6.59)$$

Noting that $\arg(x - 1) = \pi$ and $\arg(x + 1) = 0$, Eq. (6.59) becomes

$$\Psi(x, 0) = c_0 - c_3\pi + c_2 \log\sqrt{(x - 1)^2} + c_4 \log\sqrt{(x + 1)^2}. \quad (6.60)$$

Fig. 6.9 Temperatures on a combination of a circle and straight line [8]

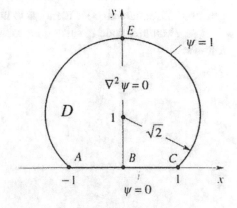

Fig. 6.10 Inscribed angle
theorem

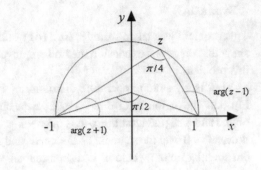

For Eq. (6.60) to vanish on the x-axis, the following must be satisfied:

$$c_2 = 0, \quad c_4 = 0, \quad c_0 = c_3\pi. \tag{6.61}$$

As shown in Fig. 6.10, from the inscribed angle theorem, we have

$$\arg(z - 1) - \arg(z + 1) = \frac{\pi}{4}. \tag{6.62}$$

Substituting Eq. (6.62) into Eq. (6.58) yields

$$
\begin{aligned}
\Psi(x, y) &= -\frac{4\arg(x + yi - 1)}{3\pi} + \frac{4\arg(x + yi + 1)}{3\pi} + \frac{4}{3} \\
&= \frac{4}{3}\left(1 - \frac{1}{\pi}(\arg(x + yi - 1) - \arg(x + yi + 1))\right) \\
&= \frac{4}{3}\left(1 - \frac{1}{\pi}\arctan\left(\frac{2y}{x^2 + y^2 - 1}\right)\right), \tag{6.63}
\end{aligned}
$$

where

$$\arg z_1 - \arg z_2 = \arg\left(\frac{z_1}{z_2}\right) \tag{6.64}$$

was used. Equation (6.63) is identical to the solution shown in [8].

A *Mathematica* code to solve this problem along with the profile of $\Psi(x, y)$ is shown as
follows:

In[ ]:= `f = (c0 + c1 I) + (c2 + c3 I) Log[z - 1] + (c4 + c5 I) Log[z + 1]`

Out[ ]= $c0 + i\,c1 + (c2 + i\,c3)\,\text{Log}[-1 + z] + (c4 + i\,c5)\,\text{Log}[1 + z]$

In[ ]:= `u = Re[f /. z → x + I y] // ComplexExpand`

Out[ ]= $c0 - c3\,\text{Arg}[-1 + x + i\,y] - c5\,\text{Arg}[1 + x + i\,y] + \frac{1}{2}\,c2\,\text{Log}\big[(-1 + x)^2 + y^2\big] + \frac{1}{2}\,c4\,\text{Log}\big[(1 + x)^2 + y^2\big]$

In[ ]:= `uxaxis = u /. y → 0`

Out[ ]= $c0 - c3\,\text{Arg}[-1 + x] - c5\,\text{Arg}[1 + x] + \frac{1}{2}\,c2\,\text{Log}\big[(-1 + x)^2\big] + \frac{1}{2}\,c4\,\text{Log}\big[(1 + x)^2\big]$

In[ ]:= `% /. {Arg[x + 1] → 0, Arg[x - 1] → Pi}`

Out[ ]= $c0 - c3\,\pi + \frac{1}{2}\,c2\,\text{Log}\big[(-1 + x)^2\big] + \frac{1}{2}\,c4\,\text{Log}\big[(1 + x)^2\big]$

In[ ]:= `u2 = u /. {c0 → c3 Pi, c2 → 0, c4 → 0, Arg[x + 1] → 0, Arg[x - 1] → Pi}`

Out[ ]= $c3\,\pi - c3\,\text{Arg}[-1 + x + i\,y] - c5\,\text{Arg}[1 + x + i\,y]$

In[ ]:= `(u2 /. Arg[x + I y - 1] → Arg[x + I y + 1] + Pi / 4) // Simplify`

Out[ ]= $\frac{3\,c3\,\pi}{4} - (c3 + c5)\,\text{Arg}[1 + x + i\,y]$

In[ ]:= `sol = Solve[{ (3 c3 π)/4 == 1, c3 + c5 == 0}, {c3, c5}][[1]]`

Out[ ]= $\left\{c3 \to \frac{}{3\,\pi},\ c5 \to -\frac{}{3\,\pi}\right\}$

In[ ]:= `psi = u2 /. sol`

Out[ ]= $\frac{4}{3} - \frac{4\,\text{Arg}[-1 + x + i\,y]}{3\,\pi} + \frac{4\,\text{Arg}[1 + x + i\,y]}{3\,\pi}$

In[ ]:= `Plot3D[psi , {x, -2, 2}, {y, 0, 5},`
`    RegionFunction → Function[{x, y, z}, x^2 + (y - 1)^2 < 2]]`

Out[ ]=

Example 4

In this example, the region is bounded by $y = 0$, $x < 0$ and $y = x$, $x > 0$ as shown in Fig. 6.11. The boundary condition is such that $T = 0$ along $y = 0$ and $T = 1$ along $y = x$.

As an analytic function, $f(z)$, whose real part represents $T(x, y)$, we assume

Fig. 6.11 Temperatures on two
straight lines

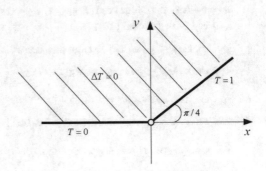

$$f(z) = (c_0 + c_1 i) + (c_2 + c_3 i) \log z, \tag{6.65}$$

as $z = 0$ is the singular point where the temperature is discontinuous. The real part of $f(z)$
is

$$T = \Re(f(z)) = c_0 + c_2 \log\left(\sqrt{x^2 + y^2}\right) - c_3 \arg z. \tag{6.66}$$

Along the negative x-axis ($y = 0$), T takes

$$T = c_0 + c_2 \log |x| - c_3 \pi, \tag{6.67}$$

from which

$$c_2 = 0, \quad c_0 = c_3 \pi. \tag{6.68}$$

For $x > 0$,

$$T = c_3 \pi - c_3 \frac{\pi}{4} = 1. \tag{6.69}$$

Therefore,

$$c_3 = \frac{4}{3\pi}. \tag{6.70}$$

Finally,

$$T = \frac{4}{3\pi} \left(\pi - \arg(x + yi)\right), \tag{6.71}$$

or

$$T = \frac{4}{3\pi} \left(\pi - \arctan\left(\frac{y}{x}\right)\right). \tag{6.72}$$

The graph for T is shown in Fig. 6.12.

Fig. 6.12 Temperatures for Example 4

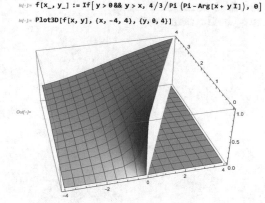

$In[\cdot]:= f[x\_, y\_] := If[y > 0 \&\& y > x, 4/3/Pi (Pi - Arg[x + y I]), 0]$

$In[\cdot]:= Plot3D[f[x, y], \{x, -4, 4\}, \{y, 0, 4\}]$

Example 5

In this example, the series expansion method is used to solve a steady-state temperature distribution problem for a two-phase material. This problem is cited in many physics and engineering textbooks on heat transfer and electromagnetism as an example of solving the Laplace equation for a two-phase system in the polar coordinate system.

As shown in Fig. 6.13, consider a 2-D circular inclusion with the thermal conductivity, k_1, and the radius, a, embedded in an infinitely extended matrix phase with the thermal conductivity, k_2. The entire medium is subject to constant heat flux (h_x, h_y) at infinity expressed as

$$k_2 \frac{\partial T}{\partial x} \to h_x, \quad k_2 \frac{\partial T}{\partial y} \to h_y \quad \text{as} \quad x, y \to \pm\infty. \tag{6.73}$$

In addition, the temperature and heat flux must be continuous at the interface of the inclusion and the matrix. Denoting the temperature distributions in the inclusion as T_1 and in the matrix as T_2, the following continuity conditions of the temperature and heat flux at $r = a$ must be satisfied as

$$T_1 = T_2 \quad \text{at} \quad r = a, \tag{6.74}$$

$$k_1 \frac{\partial T_1}{\partial n} = k_2 \frac{\partial T_2}{\partial n} \quad \text{at} \quad r = a, \tag{6.75}$$

where the directional derivative, $\partial/\partial n$, is defined as

$$\frac{\partial T}{\partial n} \equiv \frac{x}{a} \frac{\partial T}{\partial x} + \frac{y}{a} \frac{\partial T}{\partial y}. \tag{6.76}$$

As the inclusion phase has no singularity, the complex function that represents the temperature distribution inside the inclusion must be expressed by the Taylor series. On the other hand, as the matrix phase is doubly connected, the complex function that represents the temperature in the matrix phase must be expressed by the Laurent series as

Fig. 6.13 Heat conduction for
two-phase system

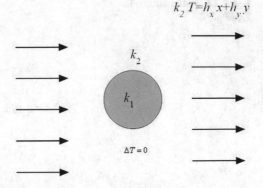

$$T_1 = \Re\left(c_0 + c_1 z + c_2 z^2\right), \tag{6.77}$$

$$T_2 = \Re\left(\frac{d_0}{z} + d_1 + d_2 z + d_3 z^2\right). \tag{6.78}$$

Substituting Eqs. (6.77) and (6.78) into Eqs. (6.74) and (6.75) and solving a set of simultaneous equations for the unknown coefficients yields

$$c_0 = c_2 = d_1 = d_3 = 0, \tag{6.79}$$

$$c_1 = \frac{2k_2}{k_1 + k_2}\left(h_x - i h_y\right), \tag{6.80}$$

$$d_0 = \frac{a^2(k_2 - k_1)}{k_1 + k_2}(h_x + i h_y), \tag{6.81}$$

$$d_2 = h_x - i h_y. \tag{6.82}$$

Therefore, the temperature distribution for each phase is expressed as

$$T_1 = \frac{2k_2}{k_1+k_2}\left(h_x x + h_y y\right), \tag{6.83}$$

$$T_2 = \left(1 - \frac{a^2(k_1-k_2)}{(k_1+k_2)}\frac{1}{x^2+y^2}\right)(h_x x + h_y y). \tag{6.84}$$

Mathematica code to solve the above problem is listed here.

```
In[•]:= z = x + I y;

In[•]:= (* t1: Temperature inside inclusion, t2: Temperature outside inclusion *)

In[•]:= t1 = Re[ (c1a + c1b I) z] // ComplexExpand;
       t2 = Re[ (d0a + d0b I) / z + (d1a + d1b I) z] // ComplexExpand;

In[•]:= (* Continuity of temperature at r=a *)

In[•]:= tempdiff = (t1 - t2) /. {x → a Cos[th], y → a Sin[th]} // Simplify // TrigReduce
```

$$Out[•]= \frac{a^2\,c1a\,Cos[th] - d0a\,Cos[th] - a^2\,d1a\,Cos[th] - a^2\,c1b\,Sin[th] - d0b\,Sin[th] + a^2\,d1b\,Sin[th]}{a}$$

```
In[•]:= (* Collect coeffs of cos(th), sin(th), cos(2h) *)

In[•]:= tempdiff2 = Collect[tempdiff, {Cos[th], Cos[2th], Sin[th]}, Simplify]
```

$$Out[•]= \left(-\frac{d0a}{a} + a\,(c1a - d1a)\right) Cos[th] + \left(-\frac{d0b}{a} + a\,(-c1b + d1b)\right) Sin[th]$$

```
In[•]:= eq1 = Coefficient[tempdiff2, Cos[th]];
       eq2 = Coefficient[tempdiff2, Sin[th]];

In[•]:= (* Continuity of heat flux at r=a *)

In[•]:= fluxdiff = k1 (x / a D[t1, x] + y / a D[t1, y]) - k2 (x / a D[t2, x] + y / a D[t2, y])
```

$$Out[•]= k1\left(\frac{c1a\,x}{a} - \frac{c1b\,y}{a}\right) - k2\left(\frac{x\left(d1a - \frac{2\,d0a\,x^2}{(x^2+y^2)^2} - \frac{2\,d0b\,x\,y}{(x^2+y^2)^2} + \frac{d0a}{x^2+y^2}\right)}{a} + \frac{y\left(-d1b - \frac{2\,d0a\,x\,y}{(x^2+y^2)^2} - \frac{2\,d0b\,y^2}{(x^2+y^2)^2} + \frac{d0b}{x^2+y^2}\right)}{a}\right)$$

```
In[•]:= fluxdiff2 = fluxdiff /. {x → a Cos[th], y → a Sin[th]} // Simplify // TrigReduce
```

$$Out[•]= \frac{1}{a^2}\left(a^2\,c1a\,k1\,Cos[th] + d0a\,k2\,Cos[th] - a^2\,d1a\,k2\,Cos[th] - a^2\,c1b\,k1\,Sin[th] + d0b\,k2\,Sin[th] + a^2\,d1b\,k2\,Sin[th]\right)$$

```
In[•]:= fluxdiff3 = Collect[fluxdiff2, {Cos[th], Sin[th]}, Simplify]
```

$$Out[•]= \left(c1a\,k1 + \frac{d0a\,k2}{a^2} - d1a\,k2\right) Cos[th] + \left(-c1b\,k1 + \left(\frac{d0b}{a^2} + d1b\right) k2\right) Sin[th]$$

```
In[•]:= eq3 = Coefficient[fluxdiff3, Cos[th]];
       eq4 = Coefficient[fluxdiff3, Sin[th]];

In[•]:= Limit[D[t2, x], {x → ∞, y → ∞}]

Out[•]= d1a

In[•]:= eq5 = d1a;

In[•]:= Limit[D[t2, y], {x → ∞, y → ∞}]

Out[•]= -d1b

In[•]:= eq6 = -d1b;

In[•]:= eqs = {eq1 == 0, eq2 == 0, eq3 == 0, eq4 == 0, eq5 == hx, eq6 == hy};

In[•]:= sol = Solve[eqs, {c1a, c1b, d0a, d0b, d1a, d1b}][[1]]
```

$$Out[•]= \left\{c1a \to \frac{2\,hx\,k2}{k1+k2},\; c1b \to -\frac{2\,hy\,k2}{k1+k2},\; d0a \to -\frac{a^2\,hx\,(k1-k2)}{k1+k2},\right.$$
$$\left. d0b \to -\frac{a^2\,hy\,(k1-k2)}{k1+k2},\; d1a \to hx,\; d1b \to -hy\right\}$$

```
In[•]:= (* temp1: Temperature inside inclusion *)

In[•]:= temp1 = t1 /. sol // Simplify
```

$$Out[•]= \frac{2\,k2\,(hx\,x + hy\,y)}{k1+k2}$$

```
In[•]:= (* temp2: Temperature outside inclusion *)

In[•]:= temp2 = t2 /. sol
```

$$Out[•]= hx\,x + hy\,y - \frac{a^2\,hx\,(k1-k2)\,x}{(k1+k2)\,(x^2+y^2)} - \frac{a^2\,hy\,(k1-k2)\,y}{(k1+k2)\,(x^2+y^2)}$$

Note that an arbitrary constant, c_0, can be added to the temperature field.

Fig. 6.14 Stress components

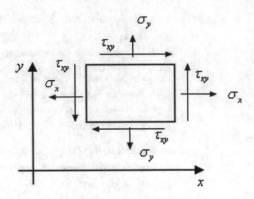

6.5   Solid Mechanics

In 2-D elasticity, three stress components, $(\sigma_x, \sigma_y, \tau_{xy})$, shown in Fig. 6.14 are the quantities that are sought from prescribed boundary conditions. In state of plane stress,[6] the balance of force equations known as the stress equilibrium equations[7] without body force are expressed as

$$\frac{\partial \sigma_x}{\partial x} + \frac{\partial \tau_{xy}}{\partial y} = 0, \tag{6.85}$$

$$\frac{\partial \tau_{yx}}{\partial x} + \frac{\partial \sigma_y}{\partial y} = 0, \tag{6.86}$$

where σ_x and σ_y are the normal components of the stress and $\tau_{xy} = \tau_{yx}$ is the shear component. The two equations of Eqs. (6.85) and (6.86) alone cannot solve for the three unknowns of σ_x, σ_y and τ_{xy} and therefore, an additional equation is needed. For an isotropic solid, the extra equation is the compatibility condition expressed in the stress components expressed as

$$\Delta \left(\sigma_x + \sigma_y \right) = 0, \tag{6.87}$$

where Δ is the 2-D Laplacian.[8]

Now that there are three equations for three unknowns, the three stress components, σ_x, σ_y and τ_{xy} can be solved from Eqs. (6.85)–(6.87).

The equilibrium equations of Eqs. (6.85) and (6.86) are automatically satisfied if all the stress components are derived from a single function, $\chi(x, y)$, as

---

[6] Plane stress is a state of stress where the normal stress component, σ_z, in the z direction is 0 exemplified by thin plates. Plane strain is a state of deformation where the normal strain component, ϵ_z, in the z direction is 0 exemplified by thick plates. A problem in plane stress can be converted to an equivalent problem in plain strain and vice versa.

[7] In 2-D, there are two equations as the balance of force in the x and y directions.

[8] The proof can be found in many books (e.g., [11]).

$$\sigma_x = \frac{\partial^2 \chi}{\partial y^2}, \quad \tau_{xy} = \tau_{yx} = -\frac{\partial^2 \chi}{\partial x \partial y}, \quad \sigma_y = \frac{\partial^2 \chi}{\partial x^2}. \tag{6.88}$$

The function, $\chi(x, y)$, is called the Airy stress function.

Substituting Eqs. (6.88) into Eq. (6.87) yields

$$\Delta^2 \chi = 0. \tag{6.89}$$

Therefore, the Airy stress function for isotropic bodies is a bi-harmonic function discussed in Sect. 6.3. The stress components, $(\sigma_x, \sigma_y, \tau_{xy})$, can be automatically derived if a bi-harmonic function, $\chi(x, y)$, is known. With Eq. (6.48), the general form of $\chi(x, y)$ is expressed as

$$\chi(x, y) = \Re \left(\bar{z} f(z) + g(z) \right), \tag{6.90}$$

where $f(z)$ and $g(z)$ are two independent analytic functions. Using Eq. (6.90), the stress components are derived directly from two independent analytic functions, $f(z)$ and $g(z)$, as[9]

$$\sigma_x = \Re \left(2f'(z) - \bar{z} f''(z) - g''(z) \right),$$
$$\sigma_y = \Re \left(2f'(z) + \bar{z} f''(z) + g''(z) \right),$$
$$\tau_{xy} = \Im \left(\bar{z} f''(z) + g''(z) \right). \tag{6.92}$$

Therefore, if analytic functions, $f(z)$ and $g(z)$, are chosen, the stress components that satisfy the equilibrium equation are derived. As for what kind of $f(z)$ and $g(z)$ one has to choose, the principle that was used for the heat conduction problems can be used, i.e., it depends on the material geometry and the boundary condition. In general, if the region considered is simply connected, $f(z)$ and $g(z)$ can be chosen as the Taylor series and if the region is multiply connected (such as holes) or has singularities such as cracks, $f(z)$ and $g(z)$ can be chosen as the Laurent series.

Computation of the stress components in Eq. (6.92) can be facilitated by the following *Mathematica* functions:

```
In[ ]:= sxx[f_, g_] :=
    (2 D[f, z] - zb D[f, {z, 2}] - D[g, {z, 2}]) /. {z → x + I y, zb → x - I y} // Re //
        ComplexExpand // Simplify
    syy[f_, g_] := (2 D[f, z] + zb D[f, {z, 2}] + D[g, {z, 2}]) /. {z → x + I y, zb → x - I y} //
        Re // ComplexExpand // Simplify
    sxy[f_, g_] := ( zb D[f, {z, 2}] + D[g, {z, 2}]) /. {z → x + I y, zb → x - I y} // Im //
        ComplexExpand // Simplify
```

---

[9] Use relationships such as

$$\frac{\partial f}{\partial x} = f'(z), \quad \frac{\partial f}{\partial y} = i f'(z), \quad \frac{\partial \bar{z}}{\partial x} = 1, \quad \frac{\partial \bar{z}}{\partial y} = -i. \tag{6.91}$$

In the code, (sxx[f,g], syy[f,g], sxy[f,g]) return $(\sigma_x, \sigma_y, \tau_{xy})$ for two analytic functions, f[z] and g[z]. For example, the stress components for

$$f(z) = z^3 + \frac{1}{z} - z, \quad g(z) = -2z^2 + 5\log(z) \tag{6.93}$$

can be computed as

In[ ]:= **f = z^3 + 1 / z - z; g = -2 z^2 + 5 Log[z];**

In[ ]:= **sxx[f, g]**

Out[ ]= $\dfrac{-5 y^4 + 2 y^6 - 12 y^8 + x^6 \left(2 - 12 y^2\right) + x^4 \left(1 + 6 y^2 - 36 y^4\right) + 6 x^2 y^2 \left(2 + y^2 - 6 y^4\right)}{\left(x^2 + y^2\right)^3}$

In[ ]:= **syy[f, g]**

Out[ ]= $\dfrac{12 x^8 + 9 y^4 - 6 y^6 + 6 x^6 \left(-1 + 6 y^2\right) + 6 x^2 y^2 \left(-2 - 3 y^2 + 2 y^4\right) + x^4 \left(-5 - 18 y^2 + 36 y^4\right)}{\left(x^2 + y^2\right)^3}$

In[ ]:= **sxy[f, g]**

Out[ ]= $\dfrac{2 x y \left(x^2 + 9 y^2\right)}{\left(x^2 + y^2\right)^3}$

In the following Examples 1–3, the media are assumed to be simply connected (no hole). Hence, $f(z)$ and $g(z)$ are expressed by the Taylor series as polynomials on z as

$$f(z) = c_1 z + c_2 z^2 + c_3 z^3, \quad g(z) = d_1 z + d_2 z^2 + d_3 z^3, \tag{6.94}$$

where $c_1 \sim d_3$ are complex unknowns. More terms can be added as necessary until the boundary conditions are satisfied.

Example 1—Uniform Stress

Consider a rectangular plate $(-a \le x \le a, -b \le y \le b)$ subject to a uniform traction force, S, along $x = \pm a$ and traction-free along $y = \pm b$ as shown in Fig. 6.15.

The boundary conditions are

$$\text{At } x = \pm a, \quad \sigma_x = S, \quad \tau_{xy} = 0, \tag{6.95}$$

$$\text{At } y = \pm b, \quad \sigma_y = 0, \quad \tau_{xy} = 0. \tag{6.96}$$

Substituting the conditions above into Eq. (6.94) determines the unknown coefficients as

$$c_1 = \frac{S}{4}, \quad c_2 = 0, \quad c_3 = 0, \quad d_1 = 0, \quad d_2 = -\frac{S}{4}, \quad d_3 = 0. \tag{6.97}$$

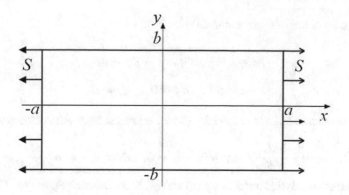

Fig. 6.15 Uniform stress

Therefore, the two analytic functions, $f(z)$ and $g(z)$, are determined as

$$f(z) = \frac{S}{4}z, \quad g(z) = -\frac{S}{4}z^2. \tag{6.98}$$

The stress components are thus derived as

$$\sigma_{xx} = S, \quad \sigma_{yy} = 0, \quad \tau_{xy} = 0. \tag{6.99}$$

This is not an exciting result and can be computed by hand easily but at least it shows an epitome of the series expansion method.

Example 2—Beam

Consider a rectangular bar ($-a \leq x \leq a$, $-b \leq y \leq b$) subject to a bending moment, M_x, along $x = \pm a$ and traction-free along $y = \pm b$ as shown in Fig. 6.16.

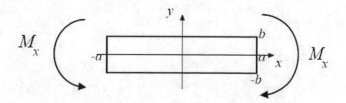

Fig. 6.16 Beam

The boundary conditions are stated as

$$\text{At } x = \pm a, \quad \int_{-b}^{b} y\, \sigma_x \, dy = M_x, \tag{6.100}$$

$$\text{At } y = \pm b, \quad \sigma_y = 0, \quad \tau_{xy} = 0. \tag{6.101}$$

Substituting the conditions above into Eq. (6.94), we can solve for the unknown coefficients as

$$c_1 = 0, \quad c_2 = -\frac{3M_x}{16b^3} i, \quad c_3 = 0, \quad d_1 = 0, \quad d_2 = 0, \quad d_3 = \frac{M_x}{16b^3} i. \tag{6.102}$$

Therefore, the two analytic functions, $f(z)$ and $g(z)$, are determined as

$$f(z) = -\frac{3M_x}{16b^3} i z^2, \quad g(z) = \frac{M_x}{16b^3} i z^3. \tag{6.103}$$

The stress components are expressed as

$$\sigma_x = \frac{3M_x\, y}{2b^3} = \frac{M}{I} y, \quad \sigma_y = 0, \quad \tau_{xy} = 0, \tag{6.104}$$

where I is the moment of inertia and is expressed as

$$I = \int_{-b}^{b} y^2 dy = \frac{2b^3}{3}. \tag{6.105}$$

The stress components, σ_x, σ_y and τ_{xy}, of Eq. (6.104) agree with those formulae found in most strength of materials textbooks (e.g., [6]).

Mathematica code to solve this beam problem is shown as follows:

```
In[ ]:= sxx[f_, g_] := (2 D[f, z] - zb D[f, {z, 2}] - D[g, {z, 2}]) /. {z → x + I y, zb → x - I y} // Re //
         ComplexExpand // Simplify
       syy[f_, g_] := (2 D[f, z] + zb D[f, {z, 2}] + D[g, {z, 2}]) /. {z → x + I y, zb → x - I y} // Re //
         ComplexExpand // Simplify
       sxy[f_, g_] := (zb D[f, {z, 2}] + D[g, {z, 2}]) /. {z → x + I y, zb → x - I y} // Im //
         ComplexExpand // Simplify
```

```
In[ ]:= (* Beam *)
```

```
In[ ]:= f = (c1 + c2 I) z + (c3 + c4 I) z^2 + (c5 + c6 I) z^3;
       g = (d1 + d2 I) z + (d3 + d4 I) z^2 + (d5 + d6 I) z^3;
```

```
In[ ]:= eq1 = (Integrate[y sxx[f, g] /. x → a, {y, -b, b}] == mx);
       eq2 = (Integrate[-y sxx[f, g] /. x → -a, {y, -b, b}] == -mx);
       j1 = syy[f, g] /. y → b;
       eq3 = Map[# == 0 &, CoefficientList[j1, x]];
       j2 = syy[f, g] /. y → -b;
       eq4 = Map[# == 0 &, CoefficientList[j2, x]];
       j3 = sxy[f, g] /. y → b;
       eq5 = Map[# == 0 &, CoefficientList[j3, x]];
       j4 = sxy[f, g] /. y → -b;
       eq6 = Map[# == 0 &, CoefficientList[j4, x]];
       alleq = {eq1, eq2, eq3, eq4, eq5, eq6} // Flatten;
```

```
In[ ]:= sol = Solve[alleq, {c1, c2, c3, c4, c5, c6, d1, d2, d3, d4, d5, d6}][[1]]
```

... Solve: Equations may not give solutions for all "solve" variables.

$$Out[ ]= \left\{ c3 \to 0, \ c4 \to -\frac{3 \, mx}{16 \, b^3}, \ c5 \to 0, \ c6 \to 0, \ d3 \to -c1, \ d4 \to 0, \ d5 \to 0, \ d6 \to \frac{mx}{16 \, b^3} \right\}$$

```
In[ ]:= ff = f /. sol
       gg = g /. sol
       sxx[ff, gg]
       syy[ff, gg]
       syy[ff, gg]
```

$$Out[ ]= \left(c1 + i \, c2\right) z - \frac{3 \, i \, mx \, z^2}{16 \, b^3}$$

$$Out[ ]= \left(d1 + i \, d2\right) z - c1 \, z^2 + \frac{i \, mx \, z^3}{16 \, b^3}$$

$$Out[ ]= 4 \, c1 + \frac{3 \, mx \, y}{2 \, b^3}$$

$$Out[ ]= 0$$

$$Out[ ]= 0$$

Example 3—Shear at End

Consider a rectangular bar ($0 \leq x \leq a$, $-b \leq y \leq b$) subject to a shear traction force, $-W$, along $x = a$ and traction-free along $y = \pm b$. The other end ($x = 0$) is fixed as shown in Fig. 6.17.

Fig. 6.17 Shear at end

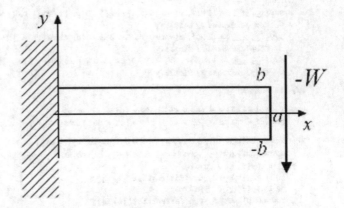

The boundary conditions are

$$\text{At } x = a, \ \sigma_x = 0, \ \int_{-b}^{b} \tau_{xy} \, dy = -W, \tag{6.106}$$

$$\text{At } y = \pm b, \ \sigma_y = 0, \ \tau_{xy} = 0. \tag{6.107}$$

Substituting the conditions above into Eq. (6.94), we can solve for the unknown coefficients as

$$c_1 = 0, \quad c_2 = -\frac{3aW}{16b^3}i, \quad c_3 = \frac{W}{16b^3}i,$$

$$d_1 = 0, \quad d_2 = -\frac{3W}{8b}i, \quad d_3 = \frac{aW}{16b^3}i, \quad d_4 = -\frac{W}{32b^3}i. \tag{6.108}$$

Therefore, the two analytic functions, $f(z)$ and $g(z)$, are determined as

$$f(z) = -\frac{3aW}{16b^3}iz^2 + \frac{W}{16b^3}iz^3, \quad g(z) = -\frac{3W}{8b}iz^2 + \frac{aW}{16b^3}iz^3 - \frac{W}{32b^3}iz^4. \tag{6.109}$$

The stress components are expressed as

$$\sigma_x = \frac{3W(a-x)y}{2b^3}, \quad \sigma_y = 0, \quad \tau_{xy} = \frac{3W(-b^2 + y^2)}{4b^3}. \tag{6.110}$$

This result agrees with those formulas found in structural mechanics textbooks (e.g., [6]).
Mathematica code to solve this beam problem is shown as follows:

```
In[ ]:= sxx[f_, g_] := (2 D[f, z] - zb D[f, {z, 2}] - D[g, {z, 2}]) /. {z → x + I y, zb → x - I y} // Re //
          ComplexExpand // Simplify
        syy[f_, g_] := (2 D[f, z] + zb D[f, {z, 2}] + D[g, {z, 2}]) /. {z → x + I y, zb → x - I y} // Re //
          ComplexExpand // Simplify
        sxy[f_, g_] := (zb D[f, {z, 2}] + D[g, {z, 2}]) /. {z → x + I y, zb → x - I y} // Im //
          ComplexExpand // Simplify

In[ ]:= f = (c1 + c2 I) z + (c3 + c4 I) z^2 + (c5 + c6 I) z^3;
        g = (d1 + d2 I) z + (d3 + d4 I) z^2 + (d5 + d6 I) z^3 + (d7 + d8 I) z^4;
        j1 = sxx[f, g] /. x → a;
        eq1 = Map[# == 0 &, CoefficientList[j1, y]];
        eq2 = (Integrate[sxy[f, g] /. x → a, {y, -b, b}] == -w);
        j2 = syy[f, g] /. y → b;
        eq3 = Map[# == 0 &, CoefficientList[j2, x]];
        j3 = syy[f, g] /. y → -b;
        eq4 = Map[# == 0 &, CoefficientList[j3, x]];
        j4 = sxy[f, g] /. y → b;
        eq5 = Map[# == 0 &, CoefficientList[j4, x]];
        j5 = sxy[f, g] /. y → -b;
        eq6 = Map[# == 0 &, CoefficientList[j5, x]];
        alleq = {eq1, eq2, eq3, eq4, eq5, eq6} // Flatten;
        sol = Solve[alleq, {c1, c2, c3, c4, c5, c6, d1, d2, d3, d4, d5, d6, d7, d8}][[1]];
        ff = f /. sol
        gg = g /. sol
        sxx[ff, gg]
        syy[ff, gg]
        sxy[ff, gg]
```

... Solve: Equations may not give solutions for all "solve" variables.

$$Out[ ]= \ i\, c2\, z - \frac{3\, i\, a\, w\, z^2}{16\, b^3} + \frac{i\, w\, z^3}{16\, b^3}$$

$$Out[ ]= \ (d1 + i\, d2)\, z - \frac{3\, i\, w\, z^2}{8\, b} + \frac{i\, a\, w\, z^3}{16\, b^3} - \frac{i\, w\, z^4}{32\, b^3}$$

$$Out[ ]= \ \frac{3\, w\, (a - x)\, y}{2\, b^3}$$

$$Out[ ]= \ 0$$

$$Out[ ]= \ -\frac{3\, w\, (b^2 - y^2)}{4\, b^3}$$

Example 4—Stress Concentration Due to a Hole

An infinitely extended elastic body with a hole of a radius, a, at the center is subject to the far field uniform stress as shown in Fig. 6.18. The boundary conditions at infinity are

$$\sigma_x \to \sigma_x^{\infty}, \quad \sigma_y \to \sigma_y^{\infty}, \quad \tau_{xy} \to \tau^{\infty} \quad \text{as} \quad x, y \to \pm\infty. \tag{6.111}$$

Also, the traction force along the perimeter of the hole is 0 which can be expressed as[10]

$$t_x = 0, \quad t_y = 0 \quad \text{along} \quad x^2 + y^2 = a^2. \tag{6.113}$$

---

[10] The quantities, t_x and t_y, are the x and y components of the traction force, respectively, and are related to the stress components as

$$t_x = \sigma_x n_x + \tau_{xy} n_y = \sigma_x \frac{x}{a} + \tau_{xy} \frac{y}{a}, \quad t_y = \tau_{xy} n_x + \sigma_y n_y = \tau_{xy} \frac{x}{a} + \sigma_y \frac{y}{a}. \tag{6.112}$$

Fig. 6.18 An infinitely
extended body with a hole

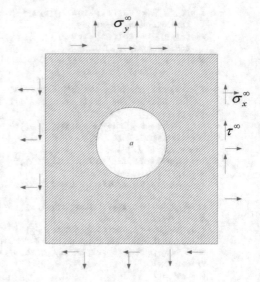

Two analytic functions, $f(z)$ and $g(z)$, are assumed to be

$$f(z) = \frac{c_{-2}}{z^2} + \frac{c_{-1}}{z} + c_0 + c_1 z, \tag{6.114}$$

$$g(z) = \frac{d_{-2}}{z^2} + \frac{d_{-1}}{z} + d_0 + d_1 z + f_1 \log z. \tag{6.115}$$

Both $f(z)$ and $g(z)$ are expressed in the Laurent series as the medium is doubly connected
(the hole). The $\log z$ term in Eq. (6.115) is added so that its derivative, $1/z$, makes $g'(z)$
complete in the Laurent series. The unknown complex coefficients, $c_{-2} \sim f_1$, are determined
to satisfy the above boundary conditions. They are solved using the *Mathematica* code shown
below as

$$c_{-2} = 0, \quad c_{-1} = \frac{a^2}{2}(\sigma_x^\infty - \sigma_y^\infty + 2i\tau^\infty), \quad c_0 = 0, \quad c_1 = \frac{\sigma_x^\infty + \sigma_y^\infty}{2},$$

$$d_{-2} = -\frac{a^4}{2}(\sigma_x^\infty - \sigma_y^\infty + 2i\tau^\infty), \quad d_{-1} = 0,$$

$$d_0 = 0, \quad d_1 = 0, \quad d_2 = \frac{1}{4}(\sigma_y^\infty - \sigma_x^\infty + 2i\tau^\infty),$$

$$f_1 = -\frac{a^2}{2}(\sigma_x^\infty + \sigma_y^\infty). \tag{6.116}$$

Therefore, $f(z)$ and $g(z)$ are determined as

$$f(z) = \frac{\frac{1}{2}a^2 \left(\sigma_x^\infty - \sigma_y^\infty\right) + ia^2\tau^\infty}{z} + z \left(\frac{\sigma_x^\infty + \sigma_y^\infty}{4}\right), \tag{6.117}$$

$$g(z) = \frac{\frac{1}{4}a^4 \left(\sigma_y^\infty - \sigma_x^\infty\right) - \frac{1}{2}ia^4\tau^\infty}{z^2} + \frac{1}{2}\log z \left(a^2(-\sigma_x^\infty) - a^2\sigma_y^\infty\right)$$
$$+ z^2 \left(\frac{\sigma_y^\infty - \sigma_x^\infty}{4} + \frac{i\tau^\infty}{2}\right). \tag{6.118}$$

The stress components are expressed as

$$\sigma_x = \frac{3a^4 \left(x^4(\sigma_x^\infty - \sigma_y^\infty) + 6x^2y^2(\sigma_y^\infty - \sigma_x^\infty) + 8\tau^\infty x^3 y - 8\tau^\infty xy^3 + y^4(\sigma_x^\infty - \sigma_y^\infty)\right)}{2\left(x^2+y^2\right)^4}$$
$$- \frac{a^2 \left(x^4(5\sigma_x^\infty - 3\sigma_y^\infty) - 12x^2y^2(\sigma_x^\infty - \sigma_y^\infty) + 24\tau^\infty x^3 y - 8\tau^\infty xy^3 - y^4(\sigma_x^\infty + \sigma_y^\infty)\right)}{2\left(x^2+y^2\right)^3} + \sigma_x^\infty,$$

$$\sigma_y = -\frac{3a^4 \left(x^4(\sigma_x^\infty - \sigma_y^\infty) + 6x^2y^2(\sigma_y^\infty - \sigma_x^\infty) + 8\tau^\infty x^3 y - 8\tau^\infty xy^3 + y^4(\sigma_x^\infty - \sigma_y^\infty)\right)}{2\left(x^2+y^2\right)^4}$$
$$+ \frac{a^2 \left(x^4(\sigma_x^\infty + \sigma_y^\infty) - 12x^2y^2(\sigma_x^\infty - \sigma_y^\infty) + 8\tau^\infty x^3 y - 24\tau^\infty xy^3 + y^4(3\sigma_x^\infty - 5\sigma_y^\infty)\right)}{2\left(x^2+y^2\right)^3} + \sigma_y^\infty,$$

$$\tau_{xy} = -\frac{3a^4 \left(\tau^\infty x^4 - 6\tau^\infty x^2 y^2 + 2x^3 y(\sigma_y^\infty - \sigma_x^\infty) + 2xy^3(\sigma_x^\infty - \sigma_y^\infty) + \tau^\infty y^4\right)}{\left(x^2+y^2\right)^4}$$
$$+ \frac{a^2 \left(2\tau^\infty x^4 - 12\tau^\infty x^2 y^2 + x^3 y(3\sigma_y^\infty - 5\sigma_x^\infty) + xy^3(3\sigma_x^\infty - 5\sigma_y^\infty) + 2\tau^\infty y^4\right)}{\left(x^2+y^2\right)^3} + \tau^\infty. \tag{6.119}$$

When $\sigma_y^\infty = \tau_{xy}^\infty = 0$, σ_x is reduced to

$$\sigma_x = \frac{\sigma_x^\infty \left(3a^4 \left(-6x^2y^2 + x^4 + y^4\right) + a^2 \left(7x^4y^2 + 13x^2y^4 - 5x^6 + y^6\right) + 2\left(x^2+y^2\right)^4\right)}{2\left(x^2+y^2\right)^4}. \tag{6.120}$$

When $x = 0$, $y = a$, the above is further reduced to

$$\sigma_x = 3\sigma_x^\infty, \tag{6.121}$$

which shows a stress concentration effect at the top edge.

A *Mathematica* code to implement the above is shown below.

```
In[ ]:= sxx[f_, g_] := (2 D[f, z] - zb D[f, {z, 2}] - D[g, {z, 2}]) /. {z → x + I y, zb → x - I y} // Re //
        ComplexExpand // Simplify
        syy[f_, g_] := (2 D[f, z] + zb D[f, {z, 2}] + D[g, {z, 2}]) /. {z → x + I y, zb → x - I y} // Re //
        ComplexExpand // Simplify
        sxy[f_, g_] := (zb D[f, {z, 2}] + D[g, {z, 2}]) /. {z → x + I y, zb → x - I y} // Im //
        ComplexExpand // Simplify

In[ ]:= (* -------------Hole----------------- *)

In[ ]:= EqMaker[f_] := Module[{j1, j2, j3, cosine, sine},
        j1 = f //. {Cos[i_. * θ] → cosine^i, Sin[j_. * θ] → sine^j};
        j2 = CoefficientList[j1, {cosine, sine}]; j3 = Flatten[j2];
        j1 = DeleteCases[j3, 0];
        Map[#1 == 0 &, j1]]

In[ ]:= fmin = -1; fmax = 1; gmin = -2; gmax = 2;

In[ ]:= f = Sum   fa[i] + I fb[i]  z^i, {i, fmin, fmax}  + f1 Log[z];
        g = Sum[ (ga[i] + I gb[i]) z^i, {i, gmin, gmax}] + g1 Log[z];
        fvar = Table[{fa[i], fb[i]}, {i, fmin, fmax}] // Flatten;
        gvar = Table[{ga[i], gb[i]}, {i, gmin, gmax}] // Flatten;
        varall = {fvar, gvar, , f1, g1} // Flatten;

In[ ]:= j1 = sxx[f, g] /. {x → r Cos[θ], y → r Sin[θ]} // Simplify // Expand;
        eq1 = {2 fa[1] - 2 ga[2] = σx∞};

In[ ]:= j2 = sxy[f, g] /. {x → r Cos[θ], y → r Sin[θ]} // Simplify // Expand;
        eq2 = {2 gb[2] = τ∞};

In[ ]:= j3 = syy[f, g] /. {x → r Cos[θ], y → r Sin[θ]} // Simplify // Expand;
        eq3 = {2 fa[1] + 2 ga[2] = σy∞};

In[ ]:= t1 = {sxx[f, g], sxy[f, g]}.{x / a, y / a} /. {x → a Cos[θ], y → a Sin[θ]} // Simplify // Expand;
        eq4 = EqMaker[t1];

In[ ]:= t2 = {sxy[f, g], syy[f, g]}.{x / a, y / a} /. {x → a Cos[θ], y → a Sin[θ]} // Simplify // Expand;
        eq5 = EqMaker[t2];

In[ ]:= eqall = {eq1, eq2, eq3, eq4, eq5} // Flatten;

In[ ]:= sol = Solve[eqall, varall][[1]]
```

... Solve: Equations may not give solutions for all "solve" variables.

$$
Out[ ]= \left\{ fa[-1] \to \frac{1}{2}\left(a^2\, \sigma x\infty - a^2\, \sigma y\infty\right),\ fb[-1] \to a^2\, \tau\infty,\ fa[1] \to \frac{\sigma x\infty + \sigma y\infty}{4}, \right.
$$
$$
ga[-2] \to \frac{1}{4}\left(-a^4\, \sigma x\infty + a^4\, \sigma y\infty\right),\ gb[-2] \to -\frac{a^4\, \tau\infty}{2},\ ga[-1] \to 0,\ gb[-1] \to 0,
$$
$$
\left. ga[2] \to \frac{1}{4}\left(-\sigma x\infty + \sigma y\infty\right),\ gb[2] \to \frac{\tau\infty}{2},\ f1 \to 0,\ g1 \to \frac{1}{2}\left(-a^2\, \sigma x\infty - a^2\, \sigma y\infty\right) \right\}
$$

```
In[ ]:= Sxx = sxx[f, g] /. sol // Simplify
```

$$
Out[ ]= \sigma x\infty + \frac{3\, a^4\left(x^4\,(\sigma x\infty - \sigma y\infty) + y^4\,(\sigma x\infty - \sigma y\infty) + 6\, x^2\, y^2\,(-\sigma x\infty + \sigma y\infty) + 8\, x^3\, y\, \tau\infty - 8\, x\, y^3\, \tau\infty\right)}{2\left(x^2 + y^2\right)^4} -
$$
$$
\frac{a^2\left(x^4\,(5\,\sigma x\infty - 3\,\sigma y\infty) - 12\, x^2\, y^2\,(\sigma x\infty - \sigma y\infty) - y^4\,(\sigma x\infty + \sigma y\infty) + 24\, x^3\, y\, \tau\infty - 8\, x\, y^3\, \tau\infty\right)}{2\left(x^2 + y^2\right)^3}
$$

```
In[ ]:= Sxy = sxy[f, g] /. sol // Simplify
```

$$
Out[ ]= \tau\infty - \frac{3\, a^4\left(2\, x\, y^3\,(\sigma x\infty - \sigma y\infty) + 2\, x^3\, y\,(-\sigma x\infty + \sigma y\infty) + x^4\, \tau\infty - 6\, x^2\, y^2\, \tau\infty + y^4\, \tau\infty\right)}{\left(x^2 + y^2\right)^4} +
$$
$$
\frac{a^2\left(x\, y^3\,(3\,\sigma x\infty - 5\,\sigma y\infty) + x^3\, y\,(-5\,\sigma x\infty + 3\,\sigma y\infty) + 2\, x^4\, \tau\infty - 12\, x^2\, y^2\, \tau\infty + 2\, y^4\, \tau\infty\right)}{\left(x^2 + y^2\right)^3}
$$

```
In[ ]:= Syy = syy[f, g] /. sol // Simplify
```

$$
Out[ ]= \sigma y\infty + \frac{a^2\left(y^4\,(3\,\sigma x\infty - 5\,\sigma y\infty) - 12\, x^2\, y^2\,(\sigma x\infty - \sigma y\infty) + x^4\,(\sigma x\infty + \sigma y\infty) + 8\, x^3\, y\, \tau\infty - 24\, x\, y^3\, \tau\infty\right)}{2\left(x^2 + y^2\right)^3} -
$$
$$
\frac{3\, a^4\left(x^4\,(\sigma x\infty - \sigma y\infty) + y^4\,(\sigma x\infty - \sigma y\infty) + 6\, x^2\, y^2\,(-\sigma x\infty + \sigma y\infty) + 8\, x^3\, y\, \tau\infty - 8\, x\, y^3\, \tau\infty\right)}{2\left(x^2 + y^2\right)^4}
$$

```
In[ ]:= f /. sol
```

$$
Out[ ]= \frac{\frac{1}{2}\left(a^2\, \sigma x\infty - a^2\, \sigma y\infty\right) + i\, a^2\, \tau\infty}{z} + fa[0] + i\, fb[0] + z\left(\frac{\sigma x\infty + \sigma y\infty}{4} + i\, fb[1]\right)
$$

```
In[ ]:= g /. sol
```

$$
Out[ ]= z^2\left(\frac{1}{4}\left(-\sigma x\infty + \sigma y\infty\right) + \frac{i\, \tau\infty}{2}\right) + \frac{\frac{1}{4}\left(-a^4\, \sigma x\infty + a^4\, \sigma y\infty\right) - \frac{1}{2}\, i\, a^4\, \tau\infty}{z^2} +
$$
$$
ga[0] + i\, gb[0] + z\,(ga[1] + i\, gb[1]) + \frac{1}{2}\left(-a^2\, \sigma x\infty - a^2\, \sigma y\infty\right) Log[z]
$$

```
In[·]:= (* 極座標 *)
       Srr = Sxx Cos[th]^2 + Syy Sin[th]^2 + 2 Sxy Sin[th] Cos[th] /.
         {x → r Cos[th], y → r Sin[th]} // FullSimplify

Out[·]=  (-a+r) (a+r) (r (σx∞+σy∞) - (3a -r  ((σx∞ -σy∞) Cos[2 th] +2 τ∞ Sin[2 th])
        ─────────────────────────────────────────────────────────────────────────
                                          2 r⁴

In[·]:= Sθθ = Sxx Sin[th]^2 + Syy Cos[th]^2 - 2 Sxy Sin[th] Cos[th] /.
         {x → r Cos[th], y → r Sin[th]} // FullSimplify

Out[·]=  r² (a² + r²  (σx∞ +σy∞) - (3 a⁴ + r⁴  ((σx∞ - σy∞) Cos[2 th] + 2 τ∞ Sin[2 th])
        ──────────────────────────────────────────────────────────────────────────────
                                          2 r⁴

In[·]:= τrθ = (Syy - Sxx) Sin[th] Cos[th] + Sxy (Cos[th]^2 - Sin[th]^2) /.
         {x → r Cos[th], y → r Sin[th]} // FullSimplify

Out[·]=  (-3 a⁴ + 2 a² r² + r⁴  (2 τ∞ Cos [2 th] + (-σx∞ + σy∞) Sin [2 th])
        ──────────────────────────────────────────────────────────────────
                                     2 r⁴
```

See [12] for detailed discussions on the Airy stress function.

6.6   Fluid Mechanics

One of the most popular applications of complex variables in mechanical/aerospace engineering is fluid mechanics. Every single analytic function represents a potential flow and since a combination of analytic functions is yet another analytic function, a combination of potential flows also represents another potential flow.

Consider a 2-D steady-state non-viscous flow where unknown quantities are the velocity components of (u, v).[11] Two equations are needed for the two unknowns, which are expressed as

$$\frac{\partial u}{\partial x} + \frac{\partial v}{\partial y} = 0, \tag{6.122}$$

$$\frac{\partial u}{\partial y} - \frac{\partial v}{\partial x} = 0. \tag{6.123}$$

Equation (6.122) is a reduced form of the equation of continuity[12] and Eq. (6.123) is the vortex-free (irrotational) condition.[13]

---

[11] In this section, we use (u, v) as the velocity components instead of the real and imaginary parts of an analytic function.

[12] The general form of the continuity equation is expressed as

$$\frac{\partial \rho}{\partial t} + \nabla \cdot (\rho \mathbf{v}) = 0, \tag{6.124}$$

where ρ is the mass density and $\mathbf{v}$ is the velocity field. For steady-state and $\rho = \text{const}$, the equation above is simplified to

$$\nabla \cdot \mathbf{v} = 0, \tag{6.125}$$

or in 2-D,

$$\frac{\partial u}{\partial x} + \frac{\partial v}{\partial y} = 0. \quad x \tag{6.126}$$

.

[13] The general form of the Navier-Stokes equation (the equation of motion) is expressed as

Equation (6.122) is satisfied if u and v are derived from a single function, $\Psi(x, y)$, as

$$\rho\frac{D\mathbf{v}}{Dt} = \nabla \cdot \sigma + \mathbf{b}, \tag{6.127}$$

where σ is the stress tensor and $\mathbf{b}$ is the body force. The D/Dt operator on the left-hand side is the material derivative (Lagrangian derivative) and is expressed as

$$\frac{D\mathbf{v}}{Dt} = \frac{\partial \mathbf{v}}{\partial t} + \mathbf{v} \cdot \nabla \mathbf{v}. \tag{6.128}$$

Using the following identity:

$$\mathbf{v} \cdot \nabla \mathbf{v} = \frac{1}{2}\nabla \left(\mathbf{v} \cdot \mathbf{v}\right) - \mathbf{v} \times \left(\nabla \times \mathbf{v}\right), \tag{6.129}$$

Equation (6.128) is written as

$$\frac{D\mathbf{v}}{Dt} = \frac{\partial \mathbf{v}}{\partial t} + \frac{1}{2}\nabla \left(\mathbf{v} \cdot \mathbf{v}\right) - \mathbf{v} \times \left(\nabla \times \mathbf{v}\right). \tag{6.130}$$

Using the following simplifications:

- Steady state:

$$\frac{\partial}{\partial t} = 0. \tag{6.131}$$

- Ideal fluid (non-viscous fluid)

$$\sigma = -p\mathbf{I}, \tag{6.132}$$

  where p is the pressure and I stands for the identity matrix.
- Body force derived from a potential

$$\mathbf{b} = -\nabla K. \tag{6.133}$$

- Pressure and density

$$P \equiv \int \frac{dp}{\rho}. \tag{6.134}$$

Equation (6.127) is simplified to

$$\rho\left(\frac{1}{2}\nabla \left(\mathbf{v} \cdot \mathbf{v}\right) - \mathbf{v} \times \left(\nabla \times \mathbf{v}\right)\right) = -\nabla P - \nabla K. \tag{6.135}$$

If the fluid is irrotational (vortex-free),

$$\nabla \times \mathbf{v} = 0, \tag{6.136}$$

or

$$\frac{\partial u}{\partial y} - \frac{\partial v}{\partial x} = 0, \tag{6.137}$$

it follows

$$\nabla \left(\frac{\mathbf{v} \cdot \mathbf{v}}{2} + K + P\right) = 0, \tag{6.138}$$

or

$$\frac{\mathbf{v} \cdot \mathbf{v}}{2} + K + P = \text{const.} \tag{6.139}$$

This is the well-known Bernoulli equation.

$$u = \frac{\partial \Psi}{\partial y} = \Psi_y$$
$$v = -\frac{\partial \Psi}{\partial x} = -\Psi_x. \tag{6.140}$$

Equation (6.123) is satisfied if u and v are derived from a single function, $\Phi(x, y)$, as

$$u = \frac{\partial \Phi}{\partial x} = \Phi_x,$$
$$v = \frac{\partial \Phi}{\partial y} = \Phi_y. \tag{6.141}$$

From Eqs. (6.140) and (6.141), it follows

$$\Phi_x = \Psi_y, \quad \Phi_y = -\Psi_x, \tag{6.142}$$

which are the Cauchy-Riemann equations in Φ and Ψ instead of u and v. Therefore, Φ and Ψ are the real part and the imaginary part of an analytic function, $f(z)$, as

$$f(z) = \Phi(x, y) + i\Psi(x, y). \tag{6.143}$$

The function, $f(z)$, is called a complex velocity potential. The two functions, $\Phi(x, y)$ and $\Psi(x, y)$, are called a velocity potential function and a stream function, respectively.

The velocity field, (u, v), satisfying both the continuity condition and the vortex-free condition can be derived by differentiating an analytic function, $f(z)$ as

$$f'(z) = u - iv, \tag{6.144}$$

i.e., u is the real part of $f'(z)$ and v is the imaginary part of $f'(z)$ with a minus sign.

Proof

$$\frac{\partial f}{\partial x} = \frac{\partial \Phi}{\partial x} + i\frac{\partial \Psi}{\partial x}$$
$$= u - iv. \tag{6.145}$$

On the other hand,

$$\frac{\partial f}{\partial x} = \frac{\partial f}{\partial z}\frac{\partial z}{\partial x} + \frac{\partial f}{\partial \bar{z}}\frac{\partial \bar{z}}{\partial x}$$
$$= \frac{\partial f}{\partial z} \times 1$$
$$= \frac{\partial f}{\partial z}, \tag{6.146}$$

as f is an analytic function, thus, free of $\bar{z}$, which leads to

$$w = f'(z) = u - iv. \tag{6.147}$$

□

Fig. 6.19 Streamlines for
$f(z) = \frac{1}{\sqrt{z}}$

In[ ]:= **flow[1 / Sqrt[z]]**

Out[ ]=

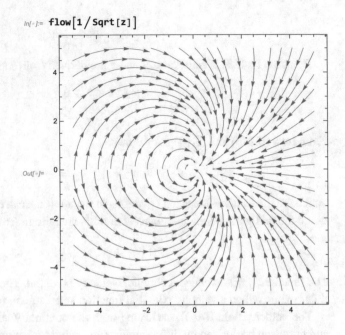

Therefore, if a complex velocity potential, $f(z)$, is known, the velocity components of the fluid can be obtained by taking the derivative of $f(z)$.

A streamline is defined as a vector field, (dx, dy), that satisfies

$$dx : dy = u : v. \tag{6.148}$$

By knowing the streamline, we can visualize the flow.

The following statement is important as we can visualize the flow by using this equation:

A set of loci, $\Psi(x, y) = const$, *forms a streamline.*

Proof Assuming $\Psi = const.$, it follows $\Psi_x dx + \Psi_y dy = 0$. Substituting Eq. (6.140) into this yields $udy - vdx = 0$, which is equivalent to $dx : dy = u : v$. $\qquad\square$

The following *Mathematica* code can be used to visualize the flow for $f(z)$.

```
In[ ]:= fluid[f_] := Module[{j1, vel, Psi}, j1 = D[f, z] /. z → x + I y;
        vel = {Re[j1], -Im[j1]} // ComplexExpand;
        Psi = Im[f /. z → x + I y] // ComplexExpand; {Psi, vel}]
     flow[f_] := StreamPlot[fluid[f][[2]], {x, -5, 5}, {y, -5, 5}]
```

In the code, the function, `flow[f[x]]`, draws the streamlines for `f[x]`. For example, Fig. 6.19 shows the streamlines for $f(z) = \frac{1}{\sqrt{z}}$.

Example 1: Uniform Flow

The simplest complex velocity potential, $f(z) = Uz$, represents a uniform flow. Using $U = U_1 + U_2 i$, we have

$$f(z) = Uz = (U_1 + U_2 i)(x + yi),$$
$$\Phi = U_1 x - U_2 y, \quad \Psi = U_2 x + U_1 y. \tag{6.149}$$

As $f'(z) = U = u - iv$, the velocity components are

$$u = U_1, \quad v = -U_2. \tag{6.150}$$

This is a uniform flow in the x direction if U is real. The streamlines are the loci of

$$\Psi = U_2 x + U_1 y = c, \tag{6.151}$$

which can be drawn for different values of c. For example, the streamlines for $f(z) = (1 - i)z$ can be drawn as (Fig. 6.20)

In[ ]:= **flow$\left[(1 - I)\ z \right]$**

Out[ ]=

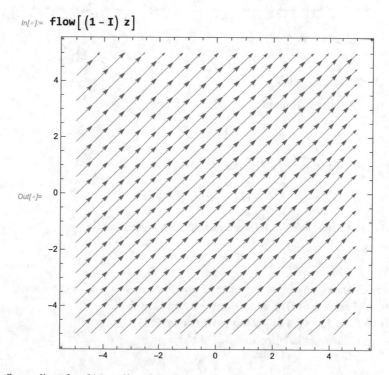

Fig. 6.20 Streamlines for $f(z) = (1 - i)z$

$In[\circ]:=$ `flow[3 Log[z]]`

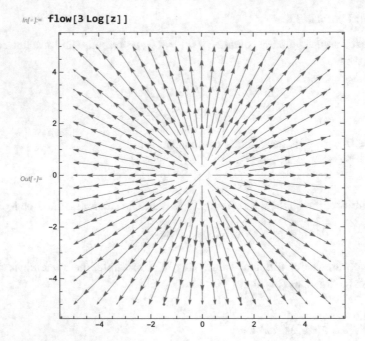

$Out[\circ]=$

Fig. 6.21 Streamlines for $f(z) = 3 \log z$

Example 2: Source and Sink

Consider the logarithmic function,

$$f(z) = m \log z. \tag{6.152}$$

By using polar form, $z = re^{i\theta}$, $f(z)$ can be written as

$$f(z) = m \left(\text{Log}\, r + i\theta\right), \tag{6.153}$$

[14]from which the equation for the streamlines is expressed as

$$\Psi = m\theta = const. \tag{6.154}$$

This flow is called a source if m is positive, and is called a sink if m is negative.

The streamlines for $f(z) = 3 \log z$ are shown in Fig. 6.21.

The streamlines for $f(z) = -2 \log z$ are shown in Fig. 6.22.

The velocity in the radial direction, v_r, is obtained as

$$v_r = \frac{\partial \Phi}{\partial r} = \frac{m}{r}. \tag{6.155}$$

---

[14] Log r is the natural logarithm for a real positive number.

$In[\circ]:=$ `flow[-2 Log[z]]`

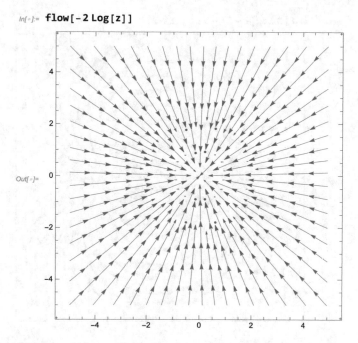

$Out[\circ]=$

Fig. 6.22 Streamlines for $f(z) = -2 \log z$

Example 3: Source and Sink Together, Doublet Flow

A combination of two potential flows is yet another potential flow because of the principle of superposition.

If a source is located at $z = a$ with a magnitude of m, and a sink is located at $z = -a$ with a magnitude of $-m$, the complex velocity potential representing such a flow is expressed as

$$f(z) = m \log (z - a) - m \log (z + a). \tag{6.156}$$

The streamlines for $f(z) = 3 \log (z - 2) - 3 \log (z + 2)$ are shown in Fig. 6.23.

It can be shown that the streamlines are part of circles. By setting

$$z - a = r_1 e^{i\theta_1}, \quad z + a = r_2 e^{i\theta_2}, \tag{6.157}$$

we have

$$
\begin{aligned}
f(z) &= m \log \left(\frac{z - a}{z + a} \right) \\
&= m \log \left(\frac{r_1}{r_2} e^{i(\theta_1 - \theta_2)} \right) \\
&= m \log \left(\frac{r_1}{r_2} \right) + m i (\theta_1 - \theta_2). \tag{6.158}
\end{aligned}
$$

In[ ]:= **flow [3 Log [z – 2] – 3 Log [z + 2]]**

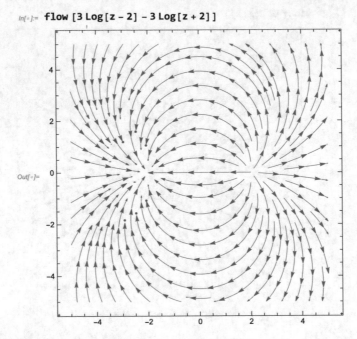

Out[ ]:=

Fig. 6.23 Streamlines for $f(z) = 3 \log (z - 2) - 3 \log (z + 2)$

Therefore, the equation for the streamlines is

$$(\theta_1 - \theta_2) = const. \tag{6.159}$$

The loci of $\theta_1 - \theta_2 = const$ form a circle as shown in Fig. 6.24.

Fig. 6.24 Streamlines as part
of a circle

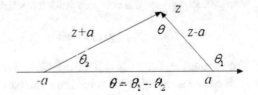

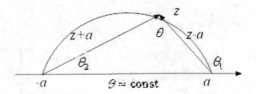

Fig. 6.25 Streamlines for
$f(z) = -2/z$

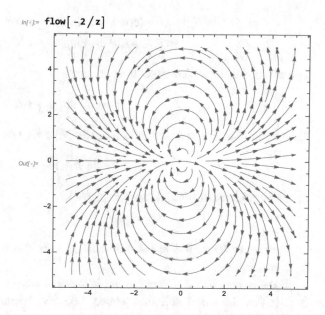

$In[\cdot]:=$ **flow$\left[\,-2\,/\,z\,\right]$**

$Out[\cdot]=$

It is interesting to see what will happen if we let the distance between the sink and source approach 0. In Eq. (6.156), let $2a \to 0$ while $2am$ remains a constant. We have

$$f(z) = m \log \left(\frac{1 - a/z}{1 + a/z} \right)$$

$$= m \left(-\frac{a}{z} - \frac{1}{2} \left(\frac{a}{z} \right)^2 - \frac{1}{3} \left(\frac{a}{z} \right)^3 - \cdots \right) - m \left(\frac{a}{z} - \frac{1}{2} \left(\frac{a}{z} \right)^2 + \frac{1}{3} \left(\frac{a}{z} \right)^3 - \cdots \right)$$

$$\sim -\frac{2am}{z} - \frac{2m}{3} \left(\frac{a}{z} \right)^3 - \frac{2m}{5} \left(\frac{a}{z} \right)^5 - \cdots$$

$$\sim -\frac{2am}{z}. \tag{6.160}$$

The flow represented by Eq. (6.160) is called a doublet flow. The streamlines for $f(z) = -2/z$ are shown in Fig. 6.25.

Example 4: Flow Around an Open Wall

As another example of the principle of superposition, consider a source with the magnitude, m, at $z = 0$ placed under a uniform flow ($= Uz$). For simplicity, U is assumed to be a real number. The complex velocity potential is expressed as

$$f(z) = Uz + m \, \log z. \tag{6.161}$$

Using polar form, $z = re^{i\theta}$, Eq. (6.161) is rewritten as

$$f(z) = Ur (\cos \theta + i \sin \theta) + m (\log r + i\theta)$$
$$= (Ur \cos \theta + m \log r) + i (Ur \sin \theta + m\theta). \qquad (6.162)$$

Therefore, the stream function, Ψ, is

$$\Psi = Ur \sin \theta + m\theta. \qquad (6.163)$$

The streamlines can be found by solving the following for r and θ:

$$Ur \sin \theta + m\theta = c. \qquad (6.164)$$

Several solutions are possible for Eq. (6.164).

1. $c = 0$
   For $c = 0$, $Ur \sin \theta + m\theta = 0$ is satisfied when $\theta = 0$ (the positive x-axis).
2. $c = m\pi$
   For $c = m\pi$, $Ur \sin \theta + m\theta = m\pi$ is satisfied when $\theta = \pi$ (the negative x-axis).
3. Besides the trivial solutions above, non-trivial solutions can be found by solving
   Eq. (6.164) for r as
   $$r = \frac{m(\pi - \theta)}{U \sin \theta}. \qquad (6.165)$$
   As $\theta \to \pi, r \to m/U$. Also, as $\theta \to 0, r \sin \theta \to m\pi/U$. For other values of θ, one can
   numerically plot r as a function of θ as Fig. 6.26.
   It is seen that the streamlines in this case separate two regions completely apart along the
   wall shown in Fig. 6.26 and the flows inside and outside the wall do not interfere with
   each other. Therefore, such a wall is considered to be a solid wall.
4. For other values of c, the streamline can be drawn numerically as in Fig. 6.27.

Equation (6.161), therefore, represents a uniform flow around an open solid wall.
Figure 6.28 is a set of streamlines derived from $f(z) = 2z + 3 \log z$.

Fig. 6.26 A solid wall that separates flow

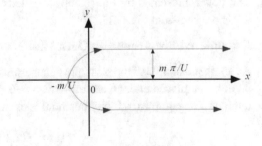

Fig. 6.27 Flow outside a solid wall

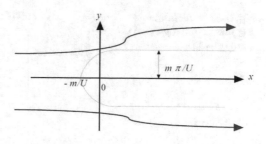

Fig. 6.28 Streamlines for $f(z) = 2z + 3\log z$

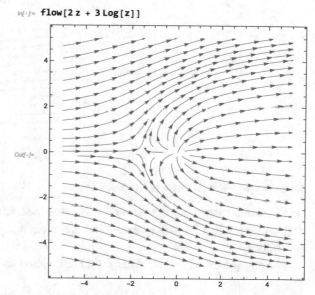

Example 5: Flow Around a Rankine Oval

If a sink with the same magnitude as in Example 4 is added at a different location, the solid wall is closed and a uniform flow around the closed solid body is realized. Thus, the complex velocity potential representing such a flow is expressed as

$$f(z) = Uz + m\log(z+a) - m\log(z-a). \tag{6.166}$$

The shape of such a closed solid body is called a Rankine oval.

Figure 6.29 is a set of streamlines of $f(z) = 3z + 2\log(z+1) - 2\log(z-1)$.

Fig. 6.29 Streamlines for
$f(z) = 3z + 2\log(z+1) - 2\log(z-1)$

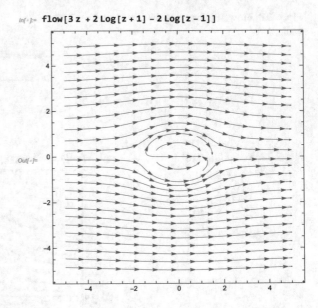

In[ ]:= `flow[3 z + 2 Log[z + 1] - 2 Log[z - 1]]`

Out[ ]=

Example 6: Flow Around a Circular Wall

As a limiting case of the flow around a Rankine oval, a flow around a circular wall can be derived by letting $a \to 0$ (i.e., a doublet). Consider a uniform flow placed on a doublet in Example 4.

The complex velocity potential, $f(z)$, is expressed as

$$f(z) = Uz + \frac{\mu}{z}. \tag{6.167}$$

With polar form, $z = re^{i\theta}$, Eq. (6.167) is written as

$$f(z) = Ure^{i\theta} + \frac{\mu}{r}e^{-i\theta}$$
$$= \left(Ur + \frac{\mu}{r}\right)\cos\theta + i\left(Ur - \frac{\mu}{r}\right)\sin\theta. \tag{6.168}$$

Therefore, the equation for the streamlines is expressed as

$$\left(Ur - \frac{\mu}{r}\right)\sin\theta = c. \tag{6.169}$$

Several solutions are possible for Eq. (6.169).

1. $c = 0$

If $\sin\theta = 0$, it follows $\theta = 0$ or π. Otherwise,

$$Ur - \mu/r = 0, \tag{6.170}$$

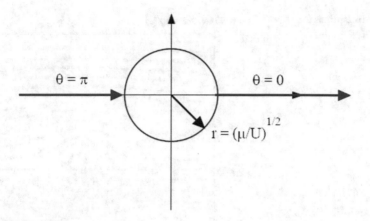

Fig. 6.30 A circle that separates flow

Fig. 6.31 Flow around a circle

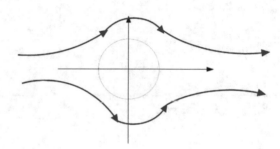

from which we have

$$r = \sqrt{\frac{\mu}{U}}. \tag{6.171}$$

This is a circle with μ/U as the radius and a solid wall that separates two regions as shown in Fig. 6.30.

2. For other values of c, the graph of the streamlines can be plotted by directly solving Eq. (6.169) as shown in Fig. 6.31.

As is seen, this represents a uniform flow around a circular cylinder.

The streamlines for $f(z) = 2z + \frac{3}{z}$ are shown in Fig. 6.32.

Fig. 6.32 Streamlines for $f(z) = 2z + \frac{3}{z}$

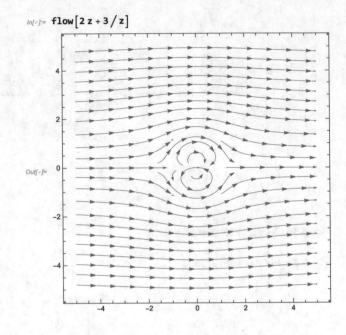

$In[\circ]:=$ **flow$\left[2\,z + 3\,/\,z\right]$**

6.7   Problems

1. Map a circle defined by $|z - 1| = 1$ in the z-plane to the w-plane by $w = \frac{1}{z}$.
2. Solve the Laplace equation ($\Delta T = 0$) shown below

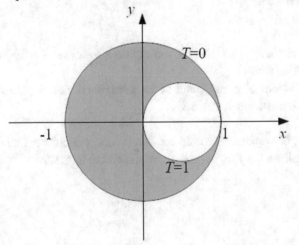

by the conformal mapping suggested as

$$w = (1 - i)\left(\frac{z - i}{z - 1}\right).$$ (6.172)

3. Solve the Laplace equation ($\Delta T = 0$) shown below by the conformal mapping of $w = z^2 + 1$:

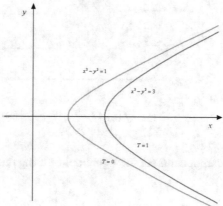

4. Solve the Laplace equation defined by the figure below

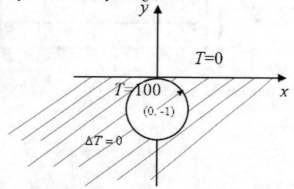

by the conformal mapping technique. Use

$$w = \frac{1}{z} \tag{6.173}$$

as a suggested mapping function.

5. Solve the same problem above but using the series expansion method. Assume

$$T = \Re\left(c_0 + \frac{c_1}{z}\right)$$

$$= \Re\left((a + bi) + \frac{c + di}{x + yi}\right), \tag{6.174}$$

and determine $a \sim d$ to satisfy the boundary conditions.

6. Find the steady-state temperature distribution outside a cylinder with a radius a in which the temperature is kept at 0 along the cylinder but subject to a constant heat flux, h, as $x, y \to \infty$:

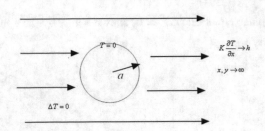

A complex function,

$$f(z) = \frac{c_0}{z} + c_1 z,$$ (6.175)

is suggested.

7. Obtain the stress components for the figure below using the Airy stress function:

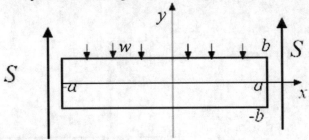

The boundary conditions are as follows:

- Along $x = a$,

$$\int_{-b}^{b} \sigma_x \, y \, dy = S.$$ (6.176)

- Along $x = -a$,

$$\int_{-b}^{b} (-\sigma_x) \, y \, dy = S.$$ (6.177)

- Along $y = b$,

$$\sigma_y = -w.$$ (6.178)

- Along $y = -b$,

$$\sigma_y = 0.$$ (6.179)

8. A complex potential function for a 2-D potential flow is given as

$$f(z) = (z + \frac{1}{z}) - \frac{i}{2\pi} \log z.$$ (6.180)

(a) Find the velocity components (u, v).
(b) Draw the streamline.

Introduction to *Mathematica*

A

The Appendix is a brief introduction to *Mathematica* and the Wolfram Language so that the reader can read *Mathematica* codes presented in the book.

Mathematica is often compared with MATLAB as both are developed for the sole purpose of solving scientific/engineering problems and both are exclusively used by engineers and scientists. However, there is a fundamental difference between *Mathematica* and MATLAB. MATLAB is primarily used for numerical computation while *Mathematica* can handle symbolic computation in addition to numerical computation. While MATLAB can use symbolic computation with Symbolic Math Toolbox, symbolic variables must be declared separately. In *Mathematica*, variables and functions can be both symbolic and numeric and they can be mixed seamlessly. It is not possible to do many of the computations discussed in this book by programming in MATLAB alone.

Many books on *Mathematica* as well as online tutorials are available although the most important reference is the one written by the creator of the software [14] which is now part of the software distribution whose entire contents can be accessed through *Mathematica*'s online help. The Wolfram Language on which *Mathematica* is written is well explained in [15].

As is with any application software, *Mathematica* has its own peculiarity and eccentricity that may take some time to get used to. Here is a list of some of the distinctive features in *Mathematica*:

- All the statements and functions are case-sensitive. The first character of all the built-in functions and constants is always capitalized.
  Examples:
  - `Sin[x]` for $\sin(x)$.
  - `Exp[x]` for $\exp(x)$.
  - `Integrate[Exp[-x] Cos[x], {x, 0, Infinity}]` for

S. Nomura, *Complex Variables for Engineers with Mathematica*, Synthesis Lectures on Mechanical Engineering, https://doi.org/10.1007/978-3-031-13067-0

$$\int_0^\infty e^{-x} \cos x \, dx.$$

  - Pi for π.
  - I for $i = \sqrt{-1}$.
- The square brackets ($[\ldots]$), not the parentheses, are used for function arguments in functions.[1]
  Examples:

  - Exp[2 + 3 I] for e^{2+3i}.
  - Log[x] for $\log x$.
- Multiplications are entered by either a * or a space.
  Examples:

  - x y for $x \times y$.
  - x*y for $x \times y$ as well.
  - xy is wrong as it represents a single variable (xy).
- Most of the built-in functions and constant names are spelled out. The first letter is capitalized as well.
  Examples:

  - Integrate[Sin[x], x] for integrations.
  - Infinity for ∞.
  - Denominator[6/7] for the denominator of 6/7.
  - N[] (Numerical) and D[] (differentiation) are some of exceptions.
- A range can be specified by the braces, {and}.
  Examples:

  - Plot[Cos[x], {x, -4, 4}] for plotting $\cos x$ over the interval between $[-4, 4]$.
  - Sum[1/n, {n, 1, 100}] for

$$\sum_{n=1}^{100} \frac{1}{n}.$$

  - Plot3D[Sin[x y], {x,-2,1},{y,-3,3}] for plotting a 3-D graph of $\sin xy$ over $-2 < x < 1$ and $-3 < y < 3$.
- Arrays and matrices are entered by (nested) braces, {and}.
  Examples:

  - vec={x, y, z} to define a vector, **vec** $= (x, y, z)$.
  - mat={{1,2,3},{4,5,6},{7,8,9}} to define a matrix,

---

[1] The reason is because f(x+y) is ambiguous as it could mean f[x+y] or f*(x+y).

$$mat = \begin{pmatrix} 1\ 2\ 3 \\ 4\ 5\ 6 \\ 7\ 8\ 9 \end{pmatrix}.$$

- A single square bracket ([···]) is used as the placeholder for function arguments while double square brackets ([[···]]) are used to refer to components in a list.
  Examples:
  - Sin[x]+Exp[-x] for $\sin(x) + \exp(-x)$.
  - m[[2, 3]] = 12 for $m_{23} = 12$.
- Free format. Ending a statement with a ";" (semicolon) suppresses output echo.
  Examples:
  - a=Expand[(x+y)^2;] assigns expansion of $(x + y)^2$ to a but the result is not displayed.
  - m[[2, 3]] = 12 sets $m_{23} = 12$ and echos back the result.

Mathematica has been available since 1988. Although many enhancements and new functionalities have been added since 1988, backward compatibility has been always maintained and most of the *Mathematica* commands that worked in an earlier version still work in the current version (Version 13 as of writing) without modifications. All of the *Mathematica* code in this book should work in any version of *Mathematica*.

Mathematica runs on many platforms including Windows, Mac and Linux with an almost identical interface throughout different environments. Although the statements and commands used in the Appendix were based on the Windows version, there should be little problem in trying those commands on different platforms.

A.1 Essential Commands/Statements

When *Mathematica* is launched first in the notebook interface, it is in input mode waiting for you to enter commands:

In[ ]:= **Expand[(x + I y) ^5]**

Out[ ]= $x^5 + 5\ i\ x^4\ y - 10\ x^3\ y^2 - 10\ i\ x^2\ y^3 + 5\ x\ y^4 + i\ y^5$

The input line from the keyboard begins with

In[1] =

and the output line from *Mathematica* begins with

Out[1] =

To recall from the preceding output result, a percentage symbol (%) can be used instead of retyping as

$In[\circ]:=$ **%**

$Out[\circ]=$ $x^5 + 5\,i\,x^4\,y - 10\,x^3\,y^2 - 10\,i\,x^2\,y^3 + 5\,x\,y^4 + i\,y^5$

The following are some examples of *Mathematica* statements that can be tried:

$In[\circ]:=$ **Integrate[x Sin[x], x]**

$Out[\circ]=$ $-x\,Cos[x] + Sin[x]$

$In[\circ]:=$ **Series[Sin[x], {x, 0, 10}]**

$Out[\circ]=$ $x - \dfrac{x^3}{6} + \dfrac{x^5}{120} - \dfrac{x^7}{5040} + \dfrac{x^9}{362880} + O[x]^{11}$

$In[\circ]:=$ **Plot$\left[$Sin[x] / x, {x, -2 Pi, 2 Pi}$\right]$**

$Out[\circ]=$

As is seen from the examples above, most of the statements are self-explanatory. After entering the statement from the keyboard, it is necessary to press the Shift and Enter keys simultaneously or the Enter key in the ten keypad for execution. Pressing the Enter key does not execute the statement but advances the cursor to the next line for additional entry of statements. All of the built-in functions in *Mathematica* such as `Integral` must be spelled out, and the first letter must be capitalized. Function parameters must be enclosed by the square brackets ([. . .]) as in `f [x]` instead of the parentheses ((. . .)). The curly brackets ({. . .}) are used to specify the range of a variable. They are also used for components of a list.

A.2   Complex Numbers

In *Mathematica*, the imaginary number, $i = \sqrt{-1}$, can be entered as `I` or by typing escape, `ii`, escape:

In[ ]:= **I**

Out[ ]= i

In[ ]:= **i^3**

Out[ ]= $-i$

In[ ]:= **I^101**

Out[ ]= i

A complex number can be entered just like any real number:

In[ ]:= **z = 2 + 3 I**

Out[ ]= $2 + 3\,i$

In[ ]:= **z^7**

Out[ ]= $6554 + 4449\,i$

In[ ]:= **Expand$\big[\,(a\,+\,b\,I)\,(c\,+\,d\,I)\,\big]$**

Out[ ]= $a\,c + i\,b\,c + i\,a\,d - b\,d$

In[ ]:= **$1\,/\,(5 + 2\,I)$**

Out[ ]= $\dfrac{5}{29} - \dfrac{2\,i}{29}$

The real part and the imaginary part of a complex number can be extracted by the `Re` and `Im` functions. The `Conjugate` function returns the complex conjugate of a complex number. The `Abs[z]` function returns the absolute value of z. The `Arg[z]` function returns the argument of z:

$In[\circ]:=$ **Re[12 – 4 I]**

$Out[\circ]=$ 12

$In[\circ]:=$ **Im[3 + 7 I]**

$Out[\circ]=$ 7

$In[\circ]:=$ **Conjugate[12 – 4 I]**

$Out[\circ]=$ 12 + 4 i

$In[\circ]:=$ **Abs[7 + 4 I]**

$Out[\circ]=$ $\sqrt{65}$

$In[\circ]:=$ **Arg[1 + Sqrt[3] I]**

$Out[\circ]=$ $\dfrac{\pi}{3}$

However, the following does not work as expected:

$In[\circ]:=$ **Re[a + b I]**

$Out[\circ]=$ $-\,\mathrm{Im}[b] + \mathrm{Re}[a]$

Because *Mathematica* assumes that all the symbolic variables (a and b) are complex numbers; it does not separate a from b. The ComplexExpand function assumes that all the symbolic variables are real:

$In[\circ]:=$ **ComplexExpand$\big[$Re$\big[$(a + b I)^5$\big]\big]$**

$Out[\circ]=$ $a^5 - 10\,a^3\,b^2 + 5\,a\,b^4$

$In[\circ]:=$ **ComplexExpand$\big[$Im$\big[$(a + b I)^5$\big]\big]$**

$Out[\circ]=$ $5\,a^4\,b - 10\,a^2\,b^3 + b^5$

If you want the numeric value of a complex number, you can use N[z] function or enter a floating number (e.g., 1.0) to force the computation to be numeric:

$In[\circ]:=$ **N[Sin[3 + 2 I]]**

$Out[\circ]=$ 0.530921 – 3.59056 i

$In[\circ]:=$ **Sin[3.0 + 2 I]**

$Out[\circ]=$ 0.530921 – 3.59056 i

$In[\circ]:=$ **Abs[Sin[3.0 + 2 I]]**

$Out[\circ]=$ 3.6296

A complex function can be visualized using the ComplexPlot3D function:

In[ ]:= `ComplexPlot3D[(z^2 + 1) / (z^2 - 1), {z, -2 - 2 I, 2 + 2 I}]`

Out[ ]=

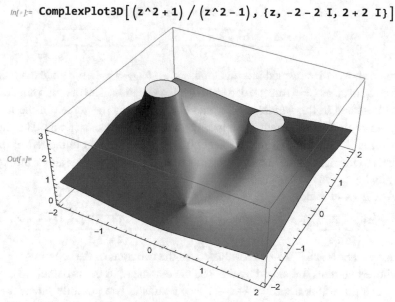

In earlier versions, you can use the `Plot3D` and `Abs` functions to achieve the same result as

In[ ]:= `Plot3D[Abs[(z^2 + 1) / (z^2 - 1) /. z → x + I y], {x, -2, 2}, {y, -2, 2}]`

Out[ ]=

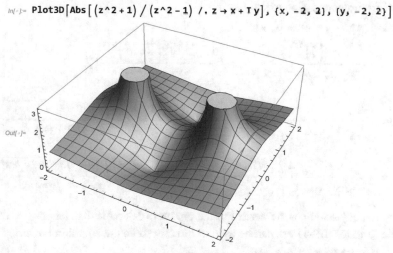

The command, `(z^2+1)/(z^2-1)/.z-> x+ I y`, substitutes $z = x + iy$ into $(z^2 + 1)/(z^2 - 1)$.

A.3 Equations

Mathematica can handle many types of equations both analytically as well as numerically. The `Solve` function and its variations can do most of the tasks as

$In[\circ]:=$ **sol = Solve[a x^2 + b x + c == 0, x]**

$Out[\circ]=$ $\left\{\left\{x \to \dfrac{-b - \sqrt{b^2 - 4\,a\,c}}{2\,a}\right\}, \left\{x \to \dfrac{-b + \sqrt{b^2 - 4\,a\,c}}{2\,a}\right\}\right\}$

The example above solves a quadratic equation, $ax^2 + bx + c = 0$, for x and outputs the two roots. Two equal signs (==) are needed to define the equation. The output from the Solve command is stored in the variable, sol, and also output to the screen in the list format with two elements each of which is enclosed by the curly brackets ({...}). The rightarrow symbol ($\to$) is the substitution rule (replace) that replaces what is to the left of the arrow with what is to the right of the arrow. Using this rule, it is possible to evaluate an expression that contains x:

$In[\circ]:=$ **ex = x^2 + x - 1 /. sol**

$Out[\circ]=$ $\left\{-1 + \dfrac{-b - \sqrt{b^2 - 4\,a\,c}}{2\,a} + \dfrac{\left(-b - \sqrt{b^2 - 4\,a\,c}\right)^2}{4\,a^2}, \ -1 + \dfrac{-b + \sqrt{b^2 - 4\,a\,c}}{2\,a} + \dfrac{\left(-b + \sqrt{b^2 - 4\,a\,c}\right)^2}{4\,a^2}\right\}$

In the example above, $x^2 + x - 1$ is evaluated for the two roots of the equation, $ax^2 + bx + c = 0$, whose two roots are stored by the Solve command to the variable, sol. As there are two roots, the evaluation of $x^2 + x - 1$ also results in two possible values for each of the roots, hence the output is in the list format.

Solve can recognize many types of equations. For an algebraic equation, it can solve up to the fifth order equation exactly as

$In[\circ]:=$ **Solve[x^3 - a x - 1 == 0, x]**

$Out[\circ]=$ $\left\{\left\{x \to \dfrac{\left(\frac{2}{3}\right)^{1/3} a}{\left(9 + \sqrt{3}\ \sqrt{27 - 4\,a^3}\right)^{1/3}} + \dfrac{\left(9 + \sqrt{3}\ \sqrt{27 - 4\,a^3}\right)^{1/3}}{2^{1/3} \times 3^{2/3}}\right\},\right.$

$\left\{x \to -\dfrac{\left(1 + i\,\sqrt{3}\right) a}{2^{2/3} \times 3^{1/3}\left(9 + \sqrt{3}\ \sqrt{27 - 4\,a^3}\right)^{1/3}} - \dfrac{\left(1 - i\,\sqrt{3}\right)\left(9 + \sqrt{3}\ \sqrt{27 - 4\,a^3}\right)^{1/3}}{2 \times 2^{1/3} \times 3^{2/3}}\right\},$

$\left.\left\{x \to -\dfrac{\left(1 - i\,\sqrt{3}\right) a}{2^{2/3} \times 3^{1/3}\left(9 + \sqrt{3}\ \sqrt{27 - 4\,a^3}\right)^{1/3}} - \dfrac{\left(1 + i\,\sqrt{3}\right)\left(9 + \sqrt{3}\ \sqrt{27 - 4\,a^3}\right)^{1/3}}{2 \times 2^{1/3} \times 3^{2/3}}\right\}\right\}$

However, in many instances, numerical values are more desirable than lengthy exact expressions. The function, NSolve, can be used to directly solve the equation numerically as

$In[\circ]:=$ **NSolve[x^3 + x - 1 == 0, x]**

$Out[\circ]=$ $\{\{x \to -0.341164 - 1.16154\ i\}, \{x \to -0.341164 + 1.16154\ i\}, \{x \to 0.682328\}\}$

A set of simultaneous equations can be solved using a list as

$In[\circ]:=$ **Solve[{2 x - y - z == 4, 4 x + y + z == 3, x - y + z == 6}, {x, y, z}]**

$Out[\circ]=$ $\left\{\left\{x \to \dfrac{7}{6},\ y \to -\dfrac{13}{4},\ z \to \dfrac{19}{12}\right\}\right\}$

The Solve command is not effective for non-linear equations. The FindRoot function can solve non-linear equations using the Newton-Raphson method as

In[ ]:= **FindRoot[Exp[-x^2] - Sin[x] == 0, {x, 2}]**

Out[ ]= $\{x \to 3.14154\}$

The statement above solves a non-linear equation,

$$e^{-x^2} - \sin x = 0,$$

with an initial guessing value of x as 2. As the Newton-Raphson method requires an initial guessing value to start, the FindRoot function uses a list, $\{x, 2\}$ as the initial value.

Differential equations (initial value problems) can be solved using DSolve command as

In[ ]:= **DSolve[y''[x] + 3 y'[x] + 2 y[x] == 3, y[x], x]**

Out[ ]= $\left\{\left\{y[x] \to \frac{3}{2} + e^{-2x} c_1 + e^{-x} c_2\right\}\right\}$

for

$$y'' + 3y' + 2y = 3. \tag{A.1}$$

The output is a rule in the list format as there may be multiple solutions available. The solution above is expressed with two integral constants, C[1] and C[2].

The DSolve function has three components. The first component is the differential equation to be solved, the second component, y[x], shows that the unknown function, y, is a function of x and the third component, x, is the variable for differentiation.

The DSolve function can also solve a set of simultaneous differential equations:

In[ ]:= **DSolve[{x'[t] - 2 y'[t] == Exp[t],**
2 x'[t] + 3 y'[t] == 1, x[0] == 1, y[0] == 2}, {x[t], y[t]}, t]

Out[ ]= $\left\{\left\{x[t] \to \frac{1}{7}\left(4 + 3 e^t + 2 t\right), y[t] \to \frac{1}{7}\left(16 - 2 e^t + t\right)\right\}\right\}$

This statement solves the following simultaneous differential equations with the initial conditions supplied:

$$x'(t) - 2y'(t) = e^t, \quad 2x'(t) + 3y'(t) = 1, \quad x(0) = 1, \quad y(0) = 2.$$

The initial conditions can be specified as part of the equation entry.

A.4 Differentiation/Integration

For differentiation of a function, the D function is used. This is one of a few built-in *Mathematica* functions that are not spelled out. Other functions which are not spelled out include Abs (the absolute value), Im (the imaginary part of a complex function) and Re (the real part of a complex function).

To differentiate $x^{100} \sin x$ with respect to x, use

In[ ]:= **D[x^100 Sin[x], x]**

Out[ ]= $x^{100} \text{Cos}[x] + 100 x^{99} \text{Sin}[x]$

Higher order derivatives can be specified by a list as (the second order derivative):

$$In[\circ]:= \textbf{D[x Exp[x], \{x, 2\}]}$$

$$Out[\circ]= 2\,e^x + e^x\,x$$

Indefinite integration of a function is evaluated by the Integrate command as

$$In[\circ]:= \textbf{Integrate[x Sin[x], x]}$$

$$Out[\circ]= -x\,Cos[x] + Sin[x]$$

To carry out definite integration of a function, it is necessary to use a list for the lower and upper bounds as

$$In[\circ]:= \textbf{Integrate[Exp[-x] Sin[x], \{x, 0, Infinity\}]}$$

$$Out[\circ]= \frac{1}{2}$$

If a function cannot be integrated analytically, it can be integrated numerically using NIntegrate as

$$In[\circ]:= \textbf{NIntegrate}\left[\, 1\,/\,(\,Cos[x]\wedge2 + 2\,),\ \{x,\ 0,\ 20\}\right]$$

$$Out[\circ]= 8.13209$$

A.5   Matrices/Vectors/Tensors

In *Mathematica*, there is no distinction between vectors and matrices. They are all represented by a list or nested lists. Hence, there is no distinction between column vectors and row vectors either.

A vector can be entered as a flat list as

$$In[\circ]:= \textbf{vec = \{v1, v2, v3\}}$$

$$Out[\circ]= \{v1,\ v2,\ v3\}$$

To make a reference to a specific component, double square brackets ([[$\cdots$]]) are used as

$$In[\circ]:= \textbf{vec[[2]]}$$

$$Out[\circ]= v2$$

The inner product between vectors can be computed by a dot (.) instead of a space or an asterisk (*) as

$$In[\circ]:= \textbf{vec.\{a[1], a[2], a[3]\}}$$

$$Out[\circ]= v1\,a[1] + v2\,a[2] + v3\,a[3]$$

A 3×3 matrix can be entered as a nested list as

$$In[\circ]:= \textbf{mat = \{\{a11, a12, a13\}, \{a21, a22, a23\}, \{a31, a32, a33\}\}}$$

$$Out[\circ]= \{\{a11,\ a12,\ a13\},\ \{a21,\ a22,\ a23\},\ \{a31,\ a32,\ a33\}\}$$

To make a reference to a specific component, double square brackets ([[$\cdots$]]) are used as

In[ ]:= **mat[[2, 3]]**

Out[ ]= **a23**

The product between a matrix and a vector or among matrices must use a dot (.) instead of a space or an asterisk (*) as

In[ ]:= **mat.vec**

Out[ ]= {a11 v1 + a12 v2 + a13 v3, a21 v1 + a22 v2 + a23 v3, a31 v1 + a32 v2 + a33 v3}

Almost all operations in linear algebra are available in *Mathematica*. This includes Inverse[mat] to inverse the matrix, mat, Eigensystem[mat] to compute the eigenvalues and the eigenvector of mat and many others. Refer to reference.

A.6   Functions

The first letter of all the built-in *Mathematica* functions is always capitalized as exemplified by Integrate, Log, Series and so on. This helps to distinguish user-defined functions from the built-in functions.

To create a user-defined function, the following syntax is used:

func[x_, y_, z_]:= definition of the function of x, y and z,

where x, y and z are the variables of the function, func. Note that an underscore follows after each variable and a colon and an equal symbol (:=) is used instead of an equal symbol. The definition of the function must be given to the right of the ":-" containing the variables with the underscores within the function argument. To define a function, $f(x) = x \sin x$, use the syntax:

In[ ]:= **f[x_] := x Sin[x]**

In[ ]:= **f[5]**

Out[ ]= **5 Sin[5]**

In[ ]:= **f[a + b]**

Out[ ]= $(a + b)$ **Sin[a + b]**

A function of several variables can be defined similarly as

In[ ]:= **f[x_, y_] := (x − y)^4**

In[ ]:= **f[2, 4]**

Out[ ]= **16**

In[ ]:= **f[5.0]**

Out[ ]= **−4.79462**

Mathematica remembers all the definitions of the function the user has input so the function above, f, returns different results depending on the number of arguments. If there is only one argument (i.e., f[5]), it returns the result of $x \sin x$ but if there are two arguments, x and y, it returns $(x - y)^4$.

If the definition of a function is more involved than a single formula requiring several steps, Module can be used to define a function that requires multiple steps with local variables as

```
In[ ]:= f[x_, y_] := Module[{a, b}, a = x + y; b = x * y; a + b]

In[ ]:= f[x, y]

Out[ ]= x + y + x y

In[ ]:= f[5, 2]

Out[ ]= 17
```

The variables, a and b, inside the braces in the Module function above are local variables inside Module that do not retain their values outside the function. Module returns the last statement, a + b, when it is called.

A.7   Graphics

Mathematica has a variety of commands that can visualize functions or data points in 2-D and 3-D. To plot a graph of $\sin x / x$ over $[-3\pi, 3\pi]$, use the Plot function as

```
In[ ]:= Plot[Sin[x] / x, {x, -3 Pi, 3 Pi}]
```

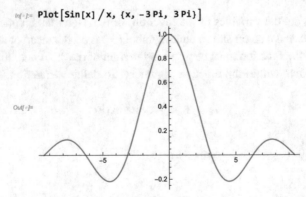

The range of the x variable can be specified by a list.

To plot multiple graphs together, use a list of functions as

In[ ]:= **Plot[{Sin[x], Cos[x]}, {x, -2 Pi, 2 Pi}]**

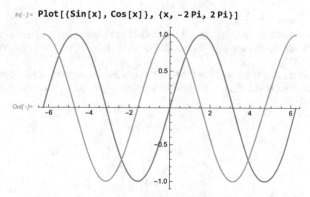

The function, Plot3D, can be used to plot a function of two variables

In[ ]:= **Plot3D[Exp[-x * y], {x, -4, 4}, {y, -5, 5}]**

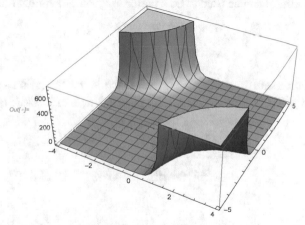

The plot range for *x* and *y* can be specified by the lists.

A.8   Other Useful Functions

Below are some of the useful functions in *Mathematica*. The usage of most of the functions is self-explanatory.

Series[f[x], {x, a, n}] returns the Taylor series of f[x] about $x = a$ up to and including the *n*-th order:

In[ ]:= **Series[Sin[x], {x, 0, 10}]**

Out[ ]= $x - \dfrac{x^3}{6} + \dfrac{x^5}{120} - \dfrac{x^7}{5040} + \dfrac{x^9}{362\,880} + O[x]^{11}$

Sum[] and NSum[] can sum up a series exactly and numerically, respectively, as

In[ ]:= **Sum[1 / i!, {i, 1, 100}]**

Out[ ]= 2 717 978 646 037 002 453 111 381 514 508 703 728 041 213 733 930 788 487 528 512 238 851 383 964 ⁞
663 670 015 423 712 227 949 076 805 203 622 200 194 928 470 090 819 123 541 234 409 774 741 771 ⁞
022 426 508 339 /
1 581 800 261 761 765 299 689 817 607 733 333 906 622 304 546 853 925 787 603 270 574 495 213 559 ⁞
207 286 705 236 295 999 595 873 191 292 435 557 980 122 436 580 528 562 896 896 000 000 000 000 ⁞
000 000 000 000

In[ ]:= **NSum[1 / i!, {i, 1, 100}]**

Out[ ]= 1.71828

Both compute

$$\sum_{i=1}^{100} \frac{1}{i!}.$$

The Apart function can be used to do a partial fraction of a rational function as

In[ ]:= **Apart[1 / (x^2 - 4)]**

Out[ ]= $\dfrac{1}{4\,(-2 + x)} - \dfrac{1}{4\,(2 + x)}$

One of the most useful functions in *Mathematica* is the Manipulate function that demonstrates the power of *Mathematica*. Here is an example of its basic usage:

In[ ]:= **Manipulate[Plot[x^a, {x, 0, 1}], {a, 1, 5}]**

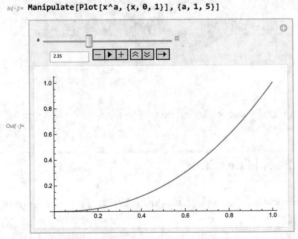

The Manipulate function works like the Table function except that the variable a in the example above can be varied by the slider in the output. By dragging the slider in the output changing the value of a, the graph of x^a is updated in real time.

A.8.1   Control Statements

Programming in *Mathematica* requires conditional statements for repetitive computations or branching out to different routines. The following are important conditional statements that are useful in *Mathematica* programming.

Do

The syntax is Do[body, range] where the contents of body are repeated for the range specified by range. The following code prints all elements of the matrix, a, defined as

$$a = \begin{pmatrix} 1\ 2\ 3 \\ 4\ 5\ 6 \\ 7\ 8\ 9 \end{pmatrix}$$

In[ ]:= **Do[Print[a[[i, j]]], {i, 3}, {j, 3}]**

1

2

3

4

5

6

7

8

9

The iteration range for the indices can be entered by the braces and if omitted, the lower bound is assumed to be 1.

Table

The syntax of the Table function is Table[expr, range] where the contents of expr are evaluated to generate a list for the range specified by range.

The Table function is used to generate a list. The following command generates a list of cos $i\pi$ when i is increased from 1 to 10:

In[ ]:= **Table[Cos[Pi j], {j, 1, 6}]**

Out[ ]= **{-1, 1, -1, 1, -1, 1}**

For

The syntax of the For function is For[init, final, incr, expr] where expr is kept evaluated while final is true with the initial counter value specified in int and the counter value is incremented by incr after each evaluation of expr similar to the For loop in C. The following code prints i from $i = 1$ to $i = 9$:

In[ ]:= `For[i = 1, i < 10, i++, Print[i]]`

1

2

3

4

5

6

7

8

9

If

The syntax of the `If` function is `If[test, body1, body2]` in which if `test` is true, `body1` is executed and if `test` is false, `body2` is executed. The following code defined a function that returns the absolute value of x (same as `Abs[x]`).

In[ ]:= `abs[x_] := If[x > 0, x, -x]`

In[ ]:= `abs[-5]`

Out[ ]= `5`

In[ ]:= `abs[1]`

Out[ ]= `1`

References

1. Lars V. Ahlfors. *Complex Analysis*. McGraw-Hill, 1979.
2. Bruce C. Berndt and Robert A. Rankin. *Ramanujan: Letters and commentary*. American Mathematical Society, 1995.
3. James Ward Brown and Ruel V. Churchill. *Complex variables and applications*. McGraw-Hill, 2009.
4. John B. Conway. *Functions of One Complex Variable I*. Springer-Verlag, 2001.
5. Robert P. Crease. *A Brief Guide to the Great Equations*. Robinson, 2009.
6. James Norman Goodier and Stephen Timoshenko. *Theory of Elasticity*. McGraw-Hill, 1970.
7. Theodore W. Gray and Jerry Glynn. *Exploring Mathematics with Mathematica*. Addison-Wesley, 1991.
8. Michael D. Greenberg. *Advanced Engineering Mathematics*. Pearson, 1998.
9. Morris Kline. Euler and infinite series. *Mathematics Magazine*, 56(5):307–315, 1983.
10. Prem K. Kythe. *Handbook of Conformal Mappings and Applications*. Chapman and Hall/CRC, 2020.
11. Seiichi Nomura. *Micromechanics with Mathematica*. John Wiley & Sons, 2016.
12. Martin H. Sadd. *Elasticity: Theory, Applications, and Numerics*. Academic Press, 2020.
13. Wikipedia. 1+2+3+, Apr. 6, 2021 [Online].
14. Stephen Wolfram. *The Mathematica Book*. Cambridge University Press, 1999.
15. Stephen Wolfram. *An Elementary Introduction to the Wolfram Language*. Wolfram Media, 2017.

© The Editor(s) (if applicable) and The Author(s), under exclusive license to Springer Nature Switzerland AG 2022
S. Nomura, *Complex Variables for Engineers with Mathematica*,
Synthesis Lectures on Mechanical Engineering,
https://doi.org/10.1007/978-3-031-13067-0

Index

A

Absolute value, 3
Airy stress function, 114, 127
Analytic, 25
Analytic continuation, 8, 53, 60, 62
Argument, 5, 7
Argument principle, 52

B

Bi-harmonic equation, 114
Bi-harmonic function, 114
Bilinear transformation, 112
Branch cut, 15
Branch point, 15

C

Cauchy-Riemann equations, 24
Cauchy's integral formula, 42, 43
Cauchy's theorem, 38
Circular inclusion, 123
Compatibility conditions, 126
Complex conjugate, 2
Complex integrals, 35
Complex plane, 4
Complex velocity potential, 139
Conformal mapping, 105
Conjugate, 2
Continuity equation, 137

D

De Moivre's formula, 7
Doublet flow, 143, 145

E

Elasticity, 126
Equation of continuity, 137
Essential singularity, 75
Euler's formula, 5
Euler's identity, 6

F

Flow around a circular wall, 148
Flow around a Rankine oval, 147
Flow around an open wall, 145
Fluid mechanics, 137
Fundamental Theorem of Algebra, 51

G

Generalized Cauchy's integral formula, 44, 45
Great Picard's theorem, 78
Green's theorem, 38

H

Harmonic function, 28, 112
Heat conduction, 116
Holomorphic, 26
Hyperbolic cosine, 10

Hyperbolic sine, 10

I
Identity theorem, 62
Imaginary axis, 4
Imaginary number, 2
Imaginary part, 2
Improper integrals, 75
Inclusion, 123
Irrotational, 137

L
Laplace equation, 28, 107, 112
Laurent series, 116
Law of exponents, 9
Liouville's theorem, 48

M
Möbius transformation, 112
Modulus, 7
Morera's theorem, 41

N
Non-viscous flow, 137

P
Picard's theorem, 78
Polar form, 7
Poles, 75
Principal value, 13

R
Rankine oval, 147
Real axis, 4

Real part, 2
Rectangular form, 7
Regular, 26
Removable singularities, 75
Residues, 75, 78
Residue theorem, 82
Riemann sphere, 83
Riemann zeta function, 64

S
Singular point, 26
Singularity, 26
Sink, 142
Solid mechanics, 126
Source, 142
Source and sink, 142
Steady-state heat conduction, 109
Stream function, 139
Streamline, 140
Stress concentration, 133
Stress equilibrium equations, 126

T
Taylor series, 53, 116

U
Uniform flow, 141

V
Velocity potential function, 139
Vortex-free, 137

W
Wolfram Cloud, 3
Wolfram Language, 153